Vascular Smooth Muscle in Culture

Volume I

Editors

Julie H. Campbell, Ph.D.
Head
Cell Biology Laboratory
Baker Medical Research Institute
Senior Research Fellow
National Health and Medical
Research Council of Australia
Melbourne, Victoria, Australia

Gordon R. Campbell, Ph.D.
Reader in Anatomy
Head
Cardiovascular Research Unit
Department of Anatomy
University of Melbourne
Melbourne, Victoria, Australia

CRC Press, Inc.
Boca Raton, Florida

Library of Congress Cataloging-in-Publication Data

Vascular smooth muscle in culture.

Includes bibliographies and index.
1. Vascular smooth muscle. 2. Tissue culture.
I. Campbell, Julie H. II. Campbell, Gordon R.
[DNLM: 1. Muscle, Smooth, Vascular--cytology. 2. Tissue Culture. WE 500 V3308]
QP110.V37V358 1987 591.1'16 86-30953
ISBN 0-8493-4325-9 (set)
ISBN 0-8493-4326-7 (v. 1)
ISBN 0-8493-4327-5 (v. 2)

Direct all inquiries to CRC Press, Inc., 2000 Corporate Blvd., N.W., Boca Raton, Florida, 33431.

International Standard Book Number 0-8493-4325-9 (Set)
International Standard Book Number 0-8493-4326-7 (Volume I)
International Standard Book Number 0-8493-4327-5 (Volume II)

Library of Congress Card Number 86-30953
Printed in the United States

PREFACE

Smooth muscle was first grown in culture in 1913/1914 by Champy,[1] Laquer[2] and Lewis and Lewis.[3] For the next 55 years, most of the reports were descriptive. However, from the late 1960s onwards, there has been an ever-increasing interest and volume of papers on cultured smooth muscle, many concerned with the vascular smooth muscle cell and its role in arterial disease. For example, in 1966, 135 papers were published on smooth muscle in culture, while in 1984, there were 847, of which 416 were specifically on vascular smooth muscle.

For these volumes we invited contributions from several of the major laboratories in the world which use cell culture as a tool to study the biology of vascular smooth muscle. These contributions cover methodology, structural and functional modulation, alterations in contractile and cytoskeletal proteins, differentiation, growth factors, growth inhibitors, collagen, glycosaminoglycan and elastin synthesis and degradation, production of prostaglandins, lipoprotein metabolism, endothelial-smooth muscle interactions, vasoactive hormone receptor expression and regulation, hypertrophy, hyperploidy, and innervation. The books thus provide a near-complete dossier of what is currently known about the vascular smooth muscle cell in culture.

Julie H. Campbell
Gordon R. Campbell
Melbourne, 1985

REFERENCES

1. **Champy, C.,** Quelques resultats de la methode de culture des tissus. I. Generalities. II. Le muscle lisse (note preliminaire), *Arch. Zool. Exp. Gen.*, 53, 42, 1913/1914.
2. **Laqueur, E.,** Zur Uberlebensdauer von Saugetierorganen mit Automatie, *Zentralbl. Physiol.*, 28, 728, 1914.
3. **Lewis, M. R. and Lewis, W. H.,** Mitochondria in tissue culture, *Science*, 39, 330, 1914.

THE EDITORS

Julie H. Campbell, Ph.D., is a Senior Research Fellow with the National Health and Medical Research Council of Australia and Head of the Cell Biology Laboratory, Baker Medical Research Institute, Prahran, Victoria, Australia.

Dr. Campbell received her B.Sc. (Hons) degree from the University of New South Wales, Sydney, Australia in 1968, and her Ph.D. degree in 1973 from the University of Melbourne, Australia. After two postdoctoral fellowships at the University of Melbourne, she was awarded the John H. Halliday Travelling Fellowship of the Life Insurance Medical Research Fund of Australia and New Zealand and further studied at University College, London, England, the University of Iowa, Iowa City and the University of Washington, Seattle, during 1976-1977. In 1978, she assumed her present position at the Baker Medical Research Institute in Melbourne, Australia.

Dr. Campbell is a member of the Australian and New Zealand Society for Cell Biology, and the Australian Society for Experimental Pathology. During 1985, she was Chairman of the Australian Atherosclerosis Group, a Committee Member for the 7th International Conference on Atherosclerosis, in Melbourne, and co-organizer for its satellite symposium "Cell Biology of the Artery Wall" held at Hamilton Island, Australia.

Dr. Campbell is the author of more than 80 papers and is coauthor of a book on cardiac muscle. Her current major research interests relate to the role of smooth muscle cells in the etiology of atherosclerosis.

Gordon R. Campbell, Ph.D., is Reader in Anatomy in the School of Medicine at the University of Melbourne, Australia. He is also Head of the Cardiovascular Research Unit within the Department of Anatomy.

Dr. Campbell received his B.Sc. (Hons) and Ph.D. degrees at the University of Melbourne in 1969 and 1974, respectively. After gaining postdoctoral experience at the University of Melbourne, University College, London, England, the University of Iowa, Iowa City and the University of Washington, Seattle, he was appointed Head, Morphology and Cell Biology Laboratory at the Baker Medical Research Institute, Prahran, Victoria, Australia in 1978. In 1980, he was appointed Senior Lecturer in the Department of Anatomy, University of Melbourne, and Reader in the same department in 1984.

Dr. Campbell is a Fellow of the Royal Society of Medicine, London, England and a member of the American Society for Cell Biology. He is also a member of the Australian and New Zealand Society for Cell Biology, the Australian Atherosclerosis Society, the Australian Society for Experimental Pathology and the Anatomical Society of Australia and New Zealand. In October 1985 he was co-organizer of an international symposium "Cell Biology of the Artery Wall" held at Hamilton Island, Australia.

Dr. Campbell is the author of more than 80 papers and the coauthor of two books on muscle structure and function. His current major research interests relate to the role of smooth muscle cells in the etiology of arteriosclerosis.

CONTRIBUTORS

R. Wayne Alexander, Ph.D., M.D.
Associate Professor of Medicine
Harvard Medical School
Cardiologist
Division of Cardiovascular Medicine
Brigham and Women's Hospital
Boston, Massachusetts

Edwin L. Bierman, M.D.
Professor of Medicine
Department of Medicine
University of Washington
Seattle, Washington

Alex Bobik, Ph.D.
Associate Director
Biochemical Pharmacology Laboratory
Alfred Hospital and Baker Medical Research Institute
Prahran, Victoria, Australia

Jacques Bonnet, M.D.
Chef de Clinique — Assistant
Department of Cardiology
Groupe Hospitalier Haut-Levêque-Xavier Arnozan
Pessac, France

Paul Bornstein, M.D.
Professor of Biochemistry and Medicine
Department of Biochemistry
University of Washington
Seattle, Washington

Daniel Bowen-Pope, Ph.D.
Research Assistant Professor
Department of Pathology
University of Washington
Seattle, Washington

H. Bricaud, M.D.
Professor of Cardiology
Chief
Department of Cardiology
Groupe Hospitalier Haut-Leveque-Xavier Arnozan
Pessac, France

John J. Castellot, Jr., Ph.D.
Assistant Professor
Department of Pathology
Harvard Medical School
Boston, Massachusetts

Alan Chait, M.D.
Professor of Medicine
Department of Medicine
University of Washington
Seattle, Washington

Peter F. Davies, Ph.D.
Associate Professor of Pathology
Department of Pathology
Brigham and Women's Hospital
Harvard Medical School
Boston, Massachusetts

B. Dorian, Ph.D.
Assistant
Unité de Recherches de Cardiologie
INSERM U8
Pessac, France

Michael A. Gimbrone, Jr., M.D.
Professor of Pathology
Brigham and Women's Hospital
Harvard Medical School
Boston, Massachusetts

David Gordon
Department of Pathology
University of Washington
Seattle, Washington

M. Daria Haust
Professor
Department of Pathology
University of Western Ontario
London, Ontario, Canada

William Hornebeck, Ph.D.
Laboratoire de Biochemie du Tissu Conjonctif
Faculte de Medicine
Universite Paris val de Marne
Creteil, France

Olli Jaakkola, M.Sc.
Assistant
Department of Biomedical Sciences
University of Tampere
Tampere, Finland

Olli-Pekka Kallioniemi, M.D.
Department of Biomedical Sciences
University of Tampere
Tampere, Finland

Morris J. Karnovsky
Professor
Department of Pathology
Harvard Medical School
Boston, Massachusetts

J. Larrue, Ph.D.
Directeur de Recherche
Unité de Recherches de Cardiologie
INSERM U8
Pessac, France

James R. A. Leushner
Department of Pathology
University of Western Ontario
London, Ontario, Canada

Richard A. Majack, Ph.D.
Research Scientist
Atherosclerosis and Thrombosis Research
The Upjohn Company
Kalamazoo, Michigan

Mark W. Majesky, Ph.D.
Postdoctoral Fellow
Department of Pathology
School of Medicine
University of Washington
Seattle, Washington

M. J. Merrilees, Ph.D.
Senior Lecturer
Department of Anatomy
University of Auckland
Auckland, New Zealand

Richard A. Murphy, Ph.D.
Professor of Physiology
Department of Physiology
University of Virginia School of Medicine
Charlottesville, Virginia

Seppo Nikkari, M.D.
Department of Biomedical Sciences
University of Tampere
Tampere, Finland

Tapio Nikkari, M.D.
Professor
Department of Biomedical Sciences
University of Tampere
Tampere, Finland

Susan E. O'Connor
Department of Pathology
Brigham and Women's Hospital
Harvard Medical School
Boston, Massachusetts

John F. Oram, Ph.D.
Research Assistant Professor
Department of Medicine
University of Washington
Seattle, Washington

Gary K. Owens, Ph.D.
Assistant Professor
Department of Physiology
University of Virginia School of Medicine
Charlottesville, Virginia

Kari Pietilä, M.D.
Special Lecturer
Department of Biomedical Science
University of Tampere
Tampere, Finland

Leslie Robert, M.D., Ph.D.
Director of Research
Laboratoire de Biochimie du Tissu Conjonctif
Faculte de Medicine
Universite Paris val de Marne
Creteil, France

Russell Ross, Ph.D.
Professor and Chairman
Department of Pathology
University of Washington
Seattle, Washington

Arthur S. Rovner, Ph.D.
Postdoctoral Fellow
Department of Physiology
University of Virginia School of Medicine
Charlottesville, Virginia

Stephen M. Schwartz, Ph.D.
Associate Professor
Department of Pathology
University of Washington
Seattle, Washington

George A. Truskey, Ph.D.
Assistant Professor of Chemical Engineering
Department of Chemical Engineering
Tufts University
Medford, Massachusetts

TABLE OF CONTENTS

Volume I

Volume II

Chapter 1

ARTERIAL SMOOTH MUSCLE DIFFERENTIATION

David Gordon and Stephen M. Schwartz

TABLE OF CONTENTS

I. INTRODUCTION

Abnormalities of vascular smooth muscle differentiation and proliferation are characteristics of both the atherosclerotic plaque and the pathologic changes of hypertensive microvessels.[1,2] Unfortunately, in the case of vascular smooth muscle cells (SMC), phenotypic states have been defined only morphologically, with hardly any biochemical characterization. Additionally, the morphological "criteria" often rest on nonspecific features, such as "numerous microfilament bundles" vs. "large amount of rough endoplasmic reticulum." Given the sampling and fixation dependence of such features, it is not surprising that there is much disagreement from one observer to the next on the state of any given cell. Finally, since these features are not really specific for SMC and since other cell types (e.g., fibroblasts, endothelial cells, and monocytes/macrophages) are known to be in the proximity of (if not within) the vessel wall, it is not always even clear whether a given cell is a "dedifferentiated" SMC or some other cell type. This is true both in vivo and in cultures derived from blood vessels.

This situation is in sharp contrast to the current understanding of the development and differentiation of striated muscle. Here the morphologic distinctions are much more sharply defined, and there has been extensive biochemical characterization of the cytoskeletal and associated proteins in relation to the different stages of proliferation and differentiation. Since the SMC has many similarities in function and cytoskeletal composition to the striated muscle, our approach is to use the knowledge already gained from skeletal muscle as an outline for investigating vascular smooth muscle proliferation and differentiation. Accordingly, this review section is organized with this analogy in mind. We begin by comparing the morphological states of skeletal vs. smooth muscle and then go on to discuss what is known about the specific muscle proteins: actin isotypes, desmin, CPK isozymes, and other cytoskeletal proteins. We end by briefly discussing atherosclerotic carotid disease, the response to endarterectomy, and suggest how a better knowledge of the controls on SMC phenotype may help in understanding this human disease and response to injury.

II. MORPHOLOGY AND CYTOSKELETAL BIOCHEMISTRY

A. Morphology

Presumptive skeletal muscle cells begin as spindle-shaped, mononucleated, diploid cells, with rather nonspecific cytoplasmic features and indeed are morphologically indistinguishable from proliferating fibroblasts.[3,4] At the time of differentiation, these precursor cells cease dividing and fuse into elongated, multinucleated "myotubes". These cells then assemble their cytoplasm into the well-known sarcomeric morphology of adult skeletal muscle.[3-5] Finally, these cells contract spontaneously in culture (in the absence of innervation), and indeed innervation is not required for the establishment of this differentiated state.[4,5]

Other differentiated features include the development of the sarcoplasmic reticulum and the T-tubule system. Each of these cells also possesses a surrounding basement membrane. The nuclei of the differentiated muscle cells are all postmitotic, diploid nuclei, and never again synthesize DNA. All muscle "regeneration" seems attributable to the proliferation and eventual fusion of undifferentiated "satellite" cells located among the well-developed myotubes.[4,6,7] The morphology of satellite cells is quite similar to that of the "presumptive myoblasts". Finally, true hypertrophy of skeletal muscle cells (cytoplasmic enlargement) can occur, for example, as a response to increased work demands. It is unclear whether or not the recruitment of satellite cells to form new myotubes contributes to this process.[8,9]

The vascular (and other) SMC are also mesenchymally derived cells and resemble fibroblasts in early development and postnatal growth.[10,11] In the developing aorta they are proliferating, spindle-shaped cells with initially much rough endoplasmic reticulum, numerous ribosomes, and otherwise nondistinguishing cytoplasmic features. During normal growth and development of the animal, there is a gradual transition to the adult morphology, which is generally a mononucleated, diploid cell whose cytoplasm is packed with actin and myosin filaments. These "myofilaments" appear randomly arranged in contrast to the highly ordered sarcomeric pattern of skeletal muscle.[11] Electron-dense "dense bodies" in the cytoplasm and adjacent to the plasma membrane are felt to be characteristic of SMC. Since actin filaments emanate from these "dense bodies", they are felt by some investigators to be the SMC counterpart to the skeletal muscle Z band.[12] The differentiated SMC also has a surrounding basement membrane and numerous pinocytotic vesicles, the latter possibly serving as an analogue to the T-tubule system of skeletal muscle. These cells can be binucleated, and with aging and hypertension, a high proportion of cells become hypertrophied with tetraploid, octaploid, or hexadecaploid content of DNA.[13-15] In this respect, the vascular SMC is more analogous to the cardiac muscle cell.[16] Finally, although some nonvascular SMC can perform spontaneous contractions in vitro,[17] vascular SMC generally require exogenous stimulation to exhibit significant contraction.[11,18,19]

In contrast to adult skeletal muscle cells, differentiated SMC do have the capacity to reenter the cell cycle in response to certain injurious stimuli,[1] and in the adult mammalian vessel wall, no morphologic subpopulation of undifferentiated "stem" or "satellite" cells have been described.[20] After injury, (e.g., balloon catheter-induced endothelial denudation of the aorta or carotid artery), the proliferating cells become morphologically indistinguishable from their less differentiated developmental precursors described above. After cessation of division, these cells resume the adult differentiated morphology.[21-23] Actually, this view is overly simplistic since one sees a whole spectrum of morphologies among these cells, with nothing approaching the clear-cut morphological changes of the skeletal muscle system. The changes are relative since each morphologic feature (e.g., dense bodies, rough endoplasmic reticulum) can be seen to some degree among the different proposed phenotypic states of adult cells.[21] Similar morphologic features can also occasionally be seen among nonmuscle cells, such as fibroblasts or endothelial cells.[11]

There is a debate over whether or not SMC showing differentiated features can also be actively proliferating at the same time. Many investigators have taken a cue from the skeletal muscle observations and have stated that SMC can be either in a "contractile" state or a proliferative/synthetic state, but not in both. The Campbells have been most vocal in this viewpoint, reporting that in vitro freshly isolated adult vascular SMC change to an undifferentiated morphology before an increase in cell number occurs (see Chapter 4). Also, the "contractile" cells do not appear to be responsive to common mitogens, such as the platelet-derived growth factor, before this morphologic transition has occurred.[21] Additionally, in those few in vitro instances in which the electron microscopic "contractile" state can be reestablished (e.g., plating at high density, co-culture with endothelium, or with heparin[21]), proliferation also ceases. They have termed this transition in phenotype "modulation"[21] as opposed to "dedifferentiation", which might imply the potential to redifferentiate also along nonsmooth muscle cell lines (something never observed). This is not a trivial theory since it implies that this phenotypic change is a requirement for SMC proliferation. Conversely, factors which might hold SMC in a differentiated state would also prevent SMC proliferation according to this theory.

However, the dichotomy between proliferation and differentiation has not been proven. Most studies in support of this theory have been done on bulk, in vitro cultures, without looking for proliferation and differentiation markers within the same individual cells. On the contrary, the Campbells have reported seeing individual SMC from vas deferens contract spontaneously, then pause for several minutes while undergoing mitosis, and then promptly resume spontaneous contractions.[24] Additionally, other investigators have reported obtaining proliferating, contractile mesenteric artery SMC in culture.[19] This suggests caution should be used in trying to predict a smooth muscle cell's functional capabilities simply from its morphological appearance. Thus, the proposed dichotomy is not absolute among SMC and may simply represent an average observation under conditions of extreme wholesale proliferation.

Again the SMC perhaps bears closer similarities to the cardiac myocytes which can possess sarcomeres and undergo mitosis simultaneously.[16,25] Through studying the expression of muscle-related proteins in various states of arterial smooth muscle, we hope to arrive at a biochemical definition of different smooth muscle phenotypic states, as well as of smooth muscle lineage, rather than relying simply on morphology. Therefore, we now turn to a discussion of muscle-specific proteins.

B. Actin Isotypes

Actin is a 42-kdalton protein existing in virtually all cell types. There are at least six distinct mammalian actin isotypes present with varying distributions among the muscle and nonmuscle tissues.[26-30] Nevertheless, amino acid sequencing has shown that this is a highly conserved class of proteins with only a small number of amino acid substitutions — indeed the same tissues from different species have much more similar actins than do different tissues within the same species. Isoelectric focusing separates the actins by isoelectric points into three general classes: α-, β-, and γ-actin.[26,29] Peptide analysis or amino acid sequencing is required to further subdivide these isotypes. Thus, adult skeletal muscle has essentially only "skeletal muscle α-actin", which differs at only four amino acid positions (out of a total of 375 residues) from "cardiac α-actin", which comprises essentially all of the actin in adult cardiac muscle cells.[29,31] These two actin isotypes are not found outside striated muscle systems. Both of these differ from the "smooth muscle α-actin", which together with "smooth muscle γ-actin", are found only in SMC. However, SMC also contain "β-actin" and "nonmuscle γ-actin", these latter two isotypes being present in most nonmuscle cells, with fibroblasts being an example.[30] Thus, as far as actin composition is concerned, the SMC appears to be somewhere between striated and nonmuscle mesenchymal cells. Although these different actin isotypes are useful for marking different tissues, the functional significance of these different isotypes is at present unknown, especially since they all can polymerize into filaments and can all interact with myosin.[31]

Within striated muscle tissues, rather striking actin isotype changes occur during normal development.[32] Thus, in skeletal muscle development, the "presumptive myoblasts" contain only β- and nonmuscle γ-actins. However, with cessation of proliferation, fusion, and during sarcomere formation, the synthesis of these actins is replaced by the production of skeletal muscle α-actin.[32,33] This may not be a simple transition, however, given observations that fetal actin mRNA from mouse skeletal muscle or from a differential skeletal muscle cell line is very similar, if not identical to, adult cardiac muscle mRNA.[35,36] Detailed analyses of proliferation and actin isotype expression in the same cell have not been possible since only recently have actin isotype-specific antibodies been generated.[37-39]

The situation in smooth muscle is even less well characterized. Chicken gizzard actin is composed essentially of only nonmuscle β- and γ-actin (presumably smooth muscle and nonmuscle types) and during its pre- and posthatching development, the γ-actin fraction increases markedly relative to the β-fraction.[40] Adult aortic smooth muscle appears unique in having a relatively high content of smooth muscle α-actin.[30,41] Thus, in adult rat aorta, 68% of the actin is smooth muscle α-actin as opposed to 18% of the total actin in rat taenia coli.[30]

Using a purified preparation of chicken gizzard actin, we have developed a monoclonal antibody which in tissue sections specifically reacts with vascular and nonvascular SMC.[38] Neither skeletal muscle nor cardiac muscle is stained with this antibody. One-dimensional "Western" blots reveal this antibody to bind exclusively to actin from smooth muscle tissues (e.g., monkey ileum) and not to skeletal muscle actin (e.g., human or rabbit skeletal muscle). Two-dimensional "Western" blots with this antibody against the chicken gizzard actin antigen and preparations of adult rat and cultured rat aorta reveal a specificity for a common antigen in α- and γ-actin isotypes. Two-dimensional gels of the immunoprecipitate of this antibody and the purified chicken gizzard actin give the same findings. Given that this monoclonal antibody stains neither striated muscle (rich in skeletal or myocardial α-actins) nor fibroblasts (cells with significant amounts of β- and nonmuscle γ-actin), we conclude that this antibody is specific for an antigen shared between smooth muscle α- and smooth muscle γ-actin. In the rat, this antibody stains all the SMC in the carotid artery and aorta. However, in the chicken thoracic aorta, it stains only alternating layers in a pattern similar to that reported for desmin. This is consistent with the ultrastructural implication that in the chicken the "interlaminar" cells are undifferentiated cells.[42] We have also seen similar heterogeneous reactivity in the bovine aorta. While the nature of the determinant common to the two smooth muscle specific actins remains unknown, this antibody offers a unique opportunity to define the cell's differentiation state in vivo and in vitro in terms of expressions of a unique determinant.

C. Desmin Expression

Desmin is one of a class of proteins called intermediate filament proteins — so named because they have diameters in the range of 100 nm, intermediate between the "thin" actin filaments and the "thick" myosin filaments.[43,44] Monomeric desmin is an approximately 52-kdalton protein present only in differentiated, striated, and smooth muscle tissues. Its function is unknown. However, in skeletal muscle, given its localization to the peri-Z-band areas in the form of a filamentous network seemingly interconnecting the Z-disks of adjacent sarcomeres, it may serve as the essential cytoskeletal framework holding the sarcomeres together in an organized pattern.[43]

More importantly for our purpose, the appearance of this protein during skeletal muscle development coincides with the transition from the proliferating myoblast to the postmitotic muscle cell, together with the switch in actin isotype to α-actin and the early formation of sarcomeres.[31] Thus it is a convenient biochemical marker for the "contractile" state in skeletal muscle development.

Another major intermediate filament protein, vimentin (molecular weight approximately 58 kdalton), is slightly more acidic than desmin on two-dimensional gels.[45-47] Vimentin is the predominant intermediate filament protein in the proliferating myoblasts. After sarcomere formation, desmin comprises the vast bulk, if not all, of the intermediate filament protein. Indeed there is some controversy as to whether or not vimentin is present in mature "contractile" skeletal muscle.[34,48]

Although intermediate filaments are visible in SMC,[11,49] their function is even less clear and the relation, if any, to the dense bodies has not been clarified.[50] The situation differs further in vascular SMC since these cells also contain substantial amounts of vimentin while in the "contractile" state, in contrast to the skeletal muscle situation.[41,51,52] Reports on the time course of appearance of desmin in smooth muscle development are currently lacking.

In the rat thoracic aorta and carotid arteries, only approximately one half of the SMC present in the media appear to react positively with antidesmin antibodies, whereas nearly all the cells react positively with antivimentin antibodies.[51] This suggests that at least for the large elastic arteries, the medial smooth muscle population is heterogeneous despite ultrastructural impression. As one goes through the abdominal aorta and into the small muscular arteries, an increasing proportion of the SMC stain positively for desmin,[53] suggesting that the SMC composition of an artery is dependent upon the size and possibly function of that artery (from elastic conductance vessels to more contractile resistance vessels, for example). Along these lines, it is interesting to note that in avian aortas (e.g., chicken and pigeon) the larger elastic arterial medias have two morphologically distinct cell populations in alternating layers: an ultrastructurally well-differentiated SMC and an "interlaminar" cell which resembles more closely a fibroblast with much less "myofilaments" and more prominent rough endoplasmic reticulum.[42,54] In the chicken, immunofluorescent stains for desmin and vimentin give results consistent with the presence of both desmin and vimentin in the smooth muscle-appearing cells, but only vimentin in the "interlaminar" cells.[52] It is interesting to speculate on the nature of this "interlaminar" cell, given the presence of desmin-negative cells in mammalian aortas. Could "interlaminar" cells be a type of smooth muscle myoblast? One still unresolved question in the rat aorta is whether or not those cells which stain negatively with the antidesmin antibody may still contain desmin, but in a nonfilamentous form.

Finally, using this same antidesmin antibody, Gabbiani et al.[51] have shown that after balloon catheter-induced arterial injury, the proliferating SMC which begin forming the neointima lack desmin. Thus desmin expression could be a cytoskeletal marker for the "contractile" differentiated state of smooth muscle.

D. Creatine Phosphokinase Isotypes

Creatine phosphokinase (CPK) is another muscle-related protein with an important role in the muscle cell's high-energy phosphate storage and delivery system. This enzyme is a dimer whose monomers come in three distinct isotypes separable by gel electrophoresis: a mitochondrial form, an M form, and a B form. Like the α-actin and desmin proteins, the M-type CPK is specific for muscle tissues.[55] The M and the B forms also have characteristic appearances in the course of skeletal muscle development. Thus in vivo, the proliferating myoblasts contain BB-CPK, but at the time of transition to postmitotic, sarcomeric muscle cells, this isotype is totally replaced by MM-CPK.[56] MM-CPK also appears with differentiation in vitro, although the transition is not complete.[56,57] Immunofluorescence studies using antibodies specific for the M form show that the onset of M-CPK synthesis is sharp with regard to the cell cycle since no cells are simultaneously positive for tritiated thymidine labeling and M-CPK labeling.[58] The presence of M-CPK is therefore a useful marker of the differentiated "contractile" state in skeletal muscle.

CPK isozyme patterns have been hardly studied at all in SMC. A few reports suggest that smooth muscle tissue (e.g., chicken gizzard, uterine smooth muscle) contains some M but mostly the B form.[55,59-61] Whether segregation by isotype exists on a cell-by-cell basis is not known. Kimes and Brandt[62] reported on two cell lines derived from rat thoracic aorta that were felt to be smooth muscle by ultrastructural criteria. These lines were reported to synthesize mainly the M form after cessation of cell division. Nevertheless, there is no clear description of the relationship of CPK isozyme expression to vascular SMC phenotype.

E. Other Muscle-Related Protein Changes

Changes in myosin composition have also been reported in the skeletal muscle systems.[32,63] Thus in rat and chicken skeletal muscle, the myosin heavy chain of embryonic muscle is different from that in neonatal muscle, which in turn is different from that in adult muscle. The first heavy chain transition is associated with the transition from proliferation to differentiation.[64] Similar changes among the different light chains have also been described.[32,65] Also of note is that slow and fast skeletal muscles contain different myosin isozymes,[65,66] a difference which appears to be a consequence of innervation by different neuron classes.[107] Although SMC have been reported to contain a distinct smooth muscle myosin as well as a nonmuscle myosin,[67-71] further detailed characterization with regard to phenotypic state is needed. For example, Larson et al.[70] have developed two polyclonal antimyosin heavy chain antibodies, one of which is smooth muscle specific (raised against human uterine myosin) and the other of which reacts with muscle and nonmuscle cells (raised against human platelet myosin). Yet since they precipitate the same myosin heavy chain from vascular smooth muscle, they have suggested that the smooth muscle distinctive feature may be some conformational change, rather than a separate smooth muscle isotype with a distinct amino acid sequence.

Tropomyosin is reported to be essentially absent in prefusion myoblast cultures of skeletal muscle and appears after fusion with differentiation.[33,34,72] Tropomyosin is present in smooth muscle,[73] but, again, there has been no full characterization with regard to development and proliferation.

III. HUMAN CAROTID ARTERY DISEASE AND MODELS OF ARTERIAL INJURY

Endarterectomy, the surgical removal of atheromatous material from an artery, is a common treatment for carotid artery disease. Unfortunately, as might be expected from animal studies, endarterectomy is probably itself a stimulus for smooth muscle proliferation and intimal thickening. A major objective is to correlate changes in animal model systems with observations on human tissue. Strandness has shown that after carotid artery endarterectomy, significant carotid stenosis of 50 to 100% luminal diameter recurs in 19% of patients within a mean postoperative follow-up interval of 16 months.[74] Some of these patients (approximately 4% of the total) will later require further vascular reconstruction at the same site for recurrent symptomatic disease. In contrast to identifiable risk factors for the development of the initial carotid artery disease lesion (diabetes, cigarette smoking, age),[75,76] the relationship between accepted atherosclerosis risk factors and recurrent carotid stenosis remains unclear.[74] Although some studies have reported restenosis rates of less than 5%,[77,78] the indication for repeat studies after surgery in these cases was chiefly neurologic symptoms. When all patients are studied postoperatively (usually by noninvasive means), higher recurrence rates are reported (9% in Turnipseed et al.,[79] 9.8% in Kremen et al.[80]). It is not surprising, therefore, that Zierler et al., using a more sophisticated imaging and flow analysis system, find a 19% recurrent rate.[74]

Other investigators have looked at the morphology of the original atherosclerotic plaque and the hyperplastic lesion recurring after endarterectomy.[77,81-83] These recurrent lesions have been variously likened to diffuse myointimal thickenings or to atherosclerotic plaques. Yet these have been strictly light microscopic studies without quantitation of the proliferative or differentiated state of the involved SMC. In particular, the relationship between the growth state of the SMC in the original plaque and the above-mentioned risk factors has not been studied. Secondly, the growth rate of the SMC remaining in the medial remnant (as represented by the portion of media removed at endarterectomy) and its relationship to the extent of subsequent intimal proliferation after endarterectomy have not been explored.

A. Kinetics of Smooth Muscle Proliferation Following Arterial Injury

Numerous animal studies have been conducted on the proliferative response of SMC to various forms of arterial injury.[1,22] We will focus on studies of a model developed in the adult rat carotid artery.[23,84-86] In this model, the formation of a thickened intima after balloon catheter-induced endothelial denudation is first due to SMC proliferation, causing a four- to fivefold increase in SMC number by 2 weeks. Thereafter, SMC number remains constant, with the further increase in intimal mass occurring by an increase in extracellular matrix. This intimal thickening, and at times also vasospasm, is responsible for luminal narrowing. In normal vessels, the thymidine labeling index of undisturbed arterial SMC is very low (0.06%). However, after arterial injury, the thymidine index rises to 40 to 70% in the media and intima within 2 days. As SMC migrate into the intima and the intima progressively widens, nearly all of the proliferation occurs within the few cell layers adjacent to the vessel lumen and ceases in reendothelialized regions. The surface of persistently denuded areas has the prespective labeling indices of 4 and 1% at 3 and 6 months after injury.

Only a few investigators have attempted to obtain cell growth kinetic data from human arteries. The frequency of polyploidy increases with age in arteries from humans.[15] Experimental studies by Owens and Schwartz[13,14] revealed a similar age dependence of the ploidy in rats. Rats also show a strong positive correlation between ploidy and systemic blood pressure. Studies to ascertain the relative proliferative capacities of diploid vs. polyploid SMC are currently under way. Preliminary results suggest a higher growth fraction among diploid as opposed to tetraploid cells in culture (unpublished results). We do not know as yet the significance of vascular SMC polyploidy. Our hypothesis is that smooth muscle polyploidization, like multinucleation in skeletal muscle or polyploidy in cardiac muscle, is characteristic of an increase in mass in differentiated muscle tissues. Polyploid SMC then might be expected to have a decreased ability to participate in the hyperplastic response required for formation of an atherosclerotic lesion.

The percent of SMC undergoing DNA synthesis was not estimated in any of the above in vivo studies. To this end, Spagnoli et al.,[88] using tritiated thymidine autoradiography on a limited number of human arterial specimens removed at surgery, found a very low labeling index (0.00 to 0.09%) in normal media, similar to that seen in adult experimental animals.[85] However, in segments of media underlying atherosclerotic plaques, there was a trend towards an increased labeling index (0.00 to 0.96%). These same investigators have published a similar preliminary thymidine labeling study on human atherosclerotic plaques and claim to see labeling only in "foam cells" or "monocyte-like cells" and only rarely in recognizable smooth muscle cells.[89] If true, this is surprising since the general view is that the proliferative population of the atherosclerotic plaque is derived from vessel wall SMC.[1] This study, however, was conducted only at the light microscopic level without specific smooth muscle and monocyte/macrophage markers and thus is not convincing. These investigators made no attempt to correlate their findings with the clinical progression of symptomatic disease at the sites examined. Given that monocytes/macrophages can produce factors that are mitogenic for mesenchymal cells,[90,91] the number of such cells in the lesion may be an important determinant of lesion progression.

In summary, there are scant data concerning the growth state and degree of SMC differentiation in adult human arteries. No one has studied the correlation in man between the progression of clinically significant atherosclerotic lesions and intimal SMC growth state or derived smooth muscle subpopulation. Variability in the proliferative and differentiated states of the cells in the lesion or underlying media could well be a valuable predictor of the fate of vessels following endarterectomy. Such studies using smooth muscle-specific and leukocyte-specific antibodies are currently under way in our laboratories.

B. Actin Isotypes of Vascular Smooth Muscle Cells

We have studied the actin isotype composition of adult rat aortic smooth muscle. Two-dimensional gels of both whole homogenized media and isolated cells obtained through enzymatic dissociation reveal that the bulk of the actin is composed of the α-isotype, with the remainder split between β- and γ-forms. This is in agreement with tryptic peptide analysis by Vandekerckhove and Weber[30] and protein electrophoresis.[92] Two-dimensional gels of multiply passaged rat aortic SMC in culture, however, reveal a marked decrease in the percentage of α-actin isotype, with the bulk of the actin now being composed of β- and γ-forms.[92] We have also seen a similar pattern in preparations of neonatal rat aortas. Since both the neonatal aortas and the cultured cells exhibit a substantial amount of proliferation in contrast to the adult rat aorta, it would appear that vascular SMC α-actin content is inversely associated with the proliferative state. In these samples we do not know whether the small amount of α-actin present represents a decrease in all of the cells or possibly an absence in all proliferating cells with full presence in noncycling, more differentiated cells.

C. Replication In Vitro vs. In Vivo

As already noted above, replication of SMC follows two quite distinct pathways in vivo. In hypertension, with the exception of necrotizing malignant hypertension, the major change is an orderly increase in cellular DNA without movement of the replicating cells.[93] At its extreme, this change occurs by endoreplications; that is, cells in the aorta and large vessels of hypertensive animals become polyploid.[94] This is reminiscent of the increase in DNA/cell seen in skeletal muscle, although in that case the change reflects fusion of differentiated cells with residual, undifferentiated stem cells.[3-5,7] In contrast, the characteristic proliferation of diffuse intimal thickening involves migration of SMC from the media into the intima.[95]

The stimuli for these responses remain unknown. For hypertension, at least in large vessels, replication occurs without evidence of platelet interaction either morphologically or by quantitative analysis of platelet adhesion.[96-98] This is important because of the large body of evidence implicating platelet release as the initiating stimulus for smooth muscle replication in atherosclerosis.[99] Even here, however, this answer may be oversimplified. Recently, Reidy and Silver[100] have shown that extensive areas of endothelium can be gently denuded without a smooth muscle proliferative response despite an exuberant thrombotic response. This contrasts with the usual proliferative response seen with the balloon catheter denudation technique.[99] We need to conclude that platelets can stimulate smooth muscle replication since they do so in vitro. The absence of an effect in vivo suggests that the normal SMC in vivo is refractory to the platelet-derived growth factor. Possible auxiliary factors include mitogenic material released from dying cells,[101] mitogens secreted by endothelial cells,[102] or mitogens released by SMC themselves.[103] Perhaps most attractive, however, is the idea that SMC in vivo exist in a state which is initially unresponsive to mitogens. This idea, originally proposed by the Campbells and discussed in Chapter 4 remains central and leaves us with the identity of the stimulus that modulates the SMC into a responsive state as a critical question.

Similarly, the stimulus for endoreplication in hypertension remains unknown. One of the few well established facts about chronic hypertension is the establishment of a fixed increase in peripheral resistance.[104] The increase in DNA could provide a "memory" for this phenomenon. Moreover, increase within a nondividing cell might be able to explain some of the abnormal physiology associated with "hypertensive" smooth muscle. Here, also, in vitro models may be useful. Goldberg et al.[105] have recently reported that isolated arterial SMC can be separated into ploidy classes by cell sorting. The cells are viable and maintain their ploidy distribution in vitro. In contrast to these results, we have found that cells in the newborn animal are diploid. When placed in culture, however, about 50% of the cultures rapidly and spontaneously develop a high frequency of polyploidy.[106] Thus it is likely that we will be able to use cell culture to study both the effects of polyploidy on cell function and the mechanism of polyploid formation itself.

IV. SUMMARY

SMC are highly differentiated cells characterized by a set of specific contractile proteins similar to those seen in skeletal muscle. In the latter cell type, differentiation is closely linked to loss of ability to divide. This tight linkage is absent in vascular smooth muscle cells since at least most of the SMC in the adult wall are capable of replicating in response to hypertension or atherosclerosis. We do, however, have probes for the specific contractile proteins and both animal cell culture model systems that should allow us to explain the nature of the linkage between differentiation and proliferation in the smooth muscle systems. We plan to apply the knowledge of SMC types and states of differentiation gained from such studies to studies of human arterial disease.

REFERENCES

1. Ross, R., Atherosclerosis: a problem of the biology of arterial wall cells and their interactions with blood components, *Arteriosclerosis,* 1, 293, 1981.
2. Benditt, E. P. and Gown, A. M., Atheroma: the artery wall and the environment, *Int. Rev. Exp. Pathol.,* 21, 55, 1980.
3. Lentz, T. L., Myogenesis: differentiation of skeletal muscle fibers, in *Developmental Regulation: Aspects of Cell Differentiation,* Coward, S. J., Ed., Academic Press, New York, 1973, 169.
4. Dientsman, S. R. and Holtzer, H., Myogenesis: a cell lineage interpretation, in *Cell Cycle and Cell Differentiation,* Reinert, J. and Holtzer, H., Eds., Springer-Verlag, New York, 1975, 1.
5. Konigsberg, I. R., Progression through G_1 and myogenic differentiation *in vitro,* in *Differentiation In Vitro,* Cambridge University Press, New York, 1982, 141.
6. Konigsberg, U. R., Lipton, B. H., and Konigsberg, I. R., The regenerative response of single mature muscle fibers isolated *in vitro, Dev. Biol.,* 45, 260, 1975.
7. Allbrook, D., Skeletal muscle regeneration, *Muscle Nerve,* 4, 234, 1981.
8. Seiden, D., A quantitative analysis of muscle cell changes in compensatory hypertrophy and work-induced hypertrophy, *Am. J. Anat.,* 145, 459, 1976.
9. DeGirolami, U. and Smith, T. W., Pathology of skeletal muscle diseases, *Am. J. Pathol.,* 107, 234, 1982.
10. Gerrity, R. G. and Cliff, W. J., The aortic tunica media of the developing rat. I. Quantitative stereologic and biochemical analysis, *Lab. Invest.,* 32, 585, 1975.
11. Chamley-Campbell, J., Campbell, G. R., and Ross, R., The smooth muscle cell in culture, *Physiol. Rev.,* 59, 1, 1979.
12. Bond, M. and Somlyo, A. V., Dense bodies and actin polarity in vertebrate smooth muscle, *J. Cell Biol.,* 95, 403, 1982.
13. Owens, G. K., Rabinovitch, P. S., and Schwartz, S. M., Smooth muscle cell hypertrophy versus hyperplasia in hypertension, *Proc. Natl. Acad. Sci. U.S.A.,* 78, 7759, 1981.
14. Owens, G. K. and Schwartz, S. M., Alterations in vascular smooth muscle mass in the spontaneously hypertensive rat. Role of cellular hypertrophy, hyperploidy and hyperplasia, *Circ. Res.,* 51, 280, 1982.
15. Barrett, T. B., Benditt, E. P., Owens, G. K., and Schwartz, S. M., Changes in cell ploidy with age in the human arterial smooth muscle cell, *Proc. Natl. Acad. Sci. U.S.A.,* 80, 882, 1982.
16. Manasek, F. J., Some comparative aspects of cardiac and skeletal myogenesis, in *Development Regulation: Aspects of Cell Differentiation,* Coward, S. J., Ed., Academic Press, New York, 1973, 193.
17. Chamley, J. H., Campbell, G. R., and Burnstock, G., Dedifferentiation, redifferentiation and bundle formation of smooth muscle cells in tissue culture: the influence of cell number and nerve fibres, *J. Embryol. Exp. Morphol.,* 32, 297, 1974.

18. Chamley-Campbell, J. H., Campbell, G. R., and Ross, R., Phenotype-dependent response of cultured aortic smooth muscle to serum mitogens, *J. Cell Biol.*, 89, 379, 1981.
19. Gunther, S., Alexander, R. W., Atkinson, W. J., and Gimbrone, M. A., Functional angiotensin II receptors in cultured vascular smooth muscle cells, *J. Cell Biol.*, 92, 289, 1982.
20. Pease, D. C. and Paule, W. J., Electron microscopy of elastic arteries: the thoracic aorta of the rat, *J. Ultrastruct. Res.*, 3, 469, 1960.
21. Chamley-Campbell, J. H. and Campbell, G. R., What controls smooth muscle phenotype? *Atherosclerosis*, 40, 347, 1981.
22. Fishman, J. A., Ryan, G. B., and Karnovsky, M. J., Endothelial regeneration in the rat carotid artery and the significance of endothelial denudation in the pathogenesis of myointimal thickening, *Lab. Invest.*, 32, 339, 1975.
23. Clowes, A. W., Reidy, M. A., and Clowes, M. M., Mechanisms of stenosis after arterial injury, *Lab. Invest.*, 49, 208, 1983.
24. Chamley, J. H. and Campbell, G. R., Mitosis of contractile smooth muscle cells in tissue culture, *Exp. Cell Res.*, 84, 105, 1974.
25. Goode, D., Mitosis of embryonic heart muscle cells *in vitro*. An immunofluorescence and ultrastructural study, *Cytobiologie Z. Exp. Zellforsch.*, 11, 203, 1975.
26. Garrels, J. I. and Gibson, W., Identification and characterization of multiple forms of actin, *Cell*, 9, 793, 1976.
27. Whalen, R. G., Burtler-Browne, G. S., and Gros, F., Protein synthesis and actin heterogeneity in calf muscle cells in culture, *Proc. Natl. Acad. Sci. U.S.A.*, 73, 2018, 1976.
28. Vandekerckhove, J. and Weber, K., At least six different actins are expressed in a higher mammal: an analysis based on the amino acid sequence of the amino-terminal tryptic peptide, *J. Mol. Biol.*, 126, 783, 1978.
29. Vandekerckhove, J. and Weber, K., The complete amino acid sequence of actins from bovine aorta, bovine heart, bovine fast skeletal muscle, and rabbit slow skeletal muscle, *Differentiation*, 14, 123, 1979.
30. Vandekerckhove, J. and Weber, K., Actin typing on total cellular extracts, *Eur. J. Biochem.*, 113, 595, 1981.
31. Korn, E. D., Actin polymerization and its regulation by proteins from nonmuscle cells, *Physiol. Rev.*, 62, 672, 1982.
32. Caplan, A. I., Fiszman, M. Y., and Eppenberger, H. M., Molecular and cell isoforms during development, *Science*, 221, 921, 1983.
33. Devlin, R. B. and Emerson, C. P., Coordinate regulation of contractile protein synthesis during myoblast differentiation, *Cell*, 13, 599, 1978.
34. Gard, D. L. and Lazarides, E., The synthesis and distribution of desmin and vimentin during myogenesis *in vitro*, *Cell*, 19, 263, 1980.
35. Minty, A. J., Alonso, S., Caravatti, M., and Buckingham, M. E., A fetal skeletal muscle actin mRNA in the mouse and its identity with cardiac actin mRNA, *Cell*, 30, 185, 1982.
36. Bains, W., Ponte, P., Blau, H., and Kedes, L., Cardiac actin is the major actin gene product in skeletal muscle cell differentiation *in vitro*, *Mol. Cell Biol.*, 4, 1449, 1984.
37. Bulinski, J. C., Kumar, S., Titani, K., and Hauschka, S. D., Peptide antibody specific for the amino terminus of skeletal muscle α-actin, *Proc. Natl. Acad. Sci. U.S.A.*, 80, 1506, 1983.
38. Gown, A. M., Vogel, A. M., Gordon, D., and Lu, P. L., A smooth muscle-specific monoclonal antibody recognizes smooth muscle actin isozymes, *J. Cell Biol.*, 100, 807, 1985.
39. Herman, I. M. and D'Amore, P. A., Microvascular pericytes contain muscle and nonmuscle actins, *J. Cell Biol.*, 101, 43, 1985.
40. Hirai, S. and Hirabayashi, T., Developmental change of protein constituents in chicken gizzards, *Dev. Biol.*, 97, 483, 1983.
41. Gabbiani, G., Schmid, E., Winter, S., Chaponnier, C., Chastonay, C. D., Vandekerckhove, J., Weber, K., and Franke, W. W., Vascular smooth muscle cells differ from other smooth muscle cells: predominancy of vimentin filaments and a specific α-type actin, *Proc. Natl. Acad. Sci. U.S.A.*, 78, 298, 1981.
42. Moss, N. S. and Benditt, E. P., Spontaneous and experimentally induced arterial lesion. I. An ultrastructural survey of the normal chicken aorta, *Lab. Invest.*, 22, 166, 1970.
43. Lazarides, E., Intermediate filaments as mechanical integrators of cellular space, *Nature (London)*, 283, 249, 1980.
44. Lazarides, E., Intermediate filaments: a chemically heterogeneous, developmentally regulated class of proteins, *Annu. Rev. Biochem.*, 51, 219, 1982.
45. Gard, D. L., Bell, P. B., and Lazarides, E., Coexistence of desmin and the fibroblastic intermediate filaments subunit in muscle and nonmuscle cells: identification and comparative peptide analysis, *Proc. Natl. Acad. Sci. U.S.A.*, 76, 3894, 1979.

46. Bravo, R. and Celis, J. E., Gene expression in normal and virally transformed mouse 3T3B and hamster BHK21 cells, *Exp. Cell Res.*, 127, 249, 1980.
47. Frank, E. D. and Warren, L., Aortic smooth muscle cells contain vimentin instead of desmin, *Proc. Natl. Acad. Sci. U.S.A.*, 78, 3020, 1981.
48. Bennett, G. S., Fellini, S. A., Toyama, Y., and Holtzer, H., Redistribution of intermediate filament subunits during skeletal myogenesis and maturation *in vitro*, *J. Cell Biol.*, 82, 577, 403, 1979.
49. Somlyo, A. P. and Somlyo, A. V., Ultrastructural aspects of activation and contraction of vascular smooth muscle, *Fed. Proc. Fed. Am. Soc. Exp. Biol.*, 35, 1288, 1976.
50. Cooke, P., A filamentous cytoskeleton in vertebrate smooth muscle fibers, *J. Cell Biol.*, 68, 539, 1976.
51. Gabbiani, G., Rungger-Brändle, E., de Chastonay, C., and Franke, W. W., Vimentin-containing smooth muscle cells in aortic intimal thickening after endothelial injury, *Lab. Invest.*, 47, 265, 1982.
52. Schmid, E., Osborn, M., Rungger-Brändle, E., Gabbiani, G., Weber, K., and Franke, W., Distribution of vimentin and desmin filaments in smooth muscle tissue of mammalian and avian aorta, *Exp. Cell Res.*, 137, 329, 1982.
53. Osborn, M., Caselitz, J., and Weber, K., Heterogeneity of intermediate filament expression in vascular smooth muscle: a gradient in desmin positive cells from the rat aortic arch to the level of the arteria iliaca communis, *Differentiation*, 20, 196, 1981.
54. Lauper, N. T., Unni, K. K., Kottke, B. A., and Titus, J. L., Anatomy and histology of aorta of White Carneau Pigeon, *Lab. Invest.*, 32, 536, 1975.
55. Rosenberg, U. B., Eppenberger, H. M., and Perriard, J., Occurrence of heterogenous forms of the subunits of creatine kinase in various muscle and nonmuscle tissues and their behavior during myogenesis, *Eur. J. Biochem.*, 116, 87, 1981.
56. Perriard, J., Caravatti, M., Perriard, E. R., and Eppenberger, H. M., Quantitation of creatine kinase isoenzyme transitions in differentiating chicken embryonic breast muscle and myogenic cell cultures by immunoadsorption, *Arch. Biochem. Biophys.*, 191, 90, 1978.
57. Caravatti, M., Perriard, J., and Eppenberger, H. M., Developmental regulation of creatine kinase isoenzymes in myogenic cell cultures from chicken, *J. Biol. Chem.*, 254, 1388, 1979.
58. Kligman, D. and Nameroff, M., Analysis of the myogenic lineage in chick embryos. I. Studies on the terminal cell division, *Exp. Cell Res.*, 125, 201, 1980.
59. Jockers-Wretou, E., Giebel, W., and Pfleiderer, G., Immunohistochemical localization of creatinkinase isoenzymes in human tissue, *Histochemistry*, 54, 83, 1977.
60. Hoag, G. N., Franks, C. R., Smith, C., and DeCoteau, W. E., Creatine kinase isoenzyme patterns in normal smooth muscle and smooth muscle neoplasms, *Clin. Biochem.*, 13, 149, 1980.
61. Iyengar, M. R., Fluellen, C. E., and Iyengar, C. W. L., Increased creatine kinase in the hormone-stimulated smooth muscle of the bovine uterus, *Biochem. Biophys. Res. Commun.*, 94, 948, 1980.
62. Kimes, B. W. and Brandt, B. L., Characterization of two putative smooth muscle cell lines from rat thoracic aorta, *Exp. Cell Res.*, 98, 349, 1976.
63. Whalen, R. G., Sell, S. M., Butler-Browne, G. S., Schwartz, K., Bouveret, P., and Pinset-Härström, I., Three myosin heavy-chain isozymes appear sequentially in rat muscle development, *Nature (London)*, 292, 805, 1981.
64. Gorza, L., Sartore, S., Triban, C., and Schiaffino, S., Embryonic-like myosin heavy chains in regenerating chicken muscle, *Exp. Cell Res.*, 143, 395, 1983.
65. Gauthier, G. F., Lowey, S., Benfield, P. A., and Hobbs, A. W., Distribution and properties of myosin isozymes in developing avian and mammalian skeletal muscle fibers, *J. Cell Biol.*, 92, 471, 1982.
66. Schiaffino, S., Askanas, V., Engel, W. K., Vitadello, M., and Sartore, S., Myosin isoenzymes in cultured human muscle, *Arch. Neurol. (Chicago)*, 39, 347, 1982.
67. Burridge, K., A comparison of fibroblast and smooth muscle myosins, *FEBS Lett.*, 45, 14, 1974.
68. Chamley, J. H., Campbell, G. R., McConnell, J. D., and Gröschel-Stewart, U., Comparison of vascular smooth muscle cells from adult human, monkey and rabbit in primary culture, *Cell Tissue Res.*, 177, 503, 1977.
69. Larson, D. M., Fujiwara, K., Alexander, R. W., and Gimbrone, M., Heterogeneity of myosin antigenic expression in vascular smooth muscle *in vivo*, *Lab. Invest.*, 50, 401, 1984.
70. Larson, D. M., Fujiwara, K., Alexander, R. W., and Gimbrone, M., Myosin in cultured vascular smooth muscle cells: immunofluorescence and immunochemical studies of alterations in antigenic expression, *J. Cell Biol.*, 99, 1582, 1984.
71. Longtine, J. A., Pinkus, G. S., Fujiwara, K., and Corson, J. M., Immunohistochemical localization of smooth muscle myosin in normal human tissues, *J. Histochem. Cytochem.*, 33, 179, 1985.
72. Crisona, N. J. and Strohman, R. C., Inhibition of contraction of cultured muscle fibers results in increased turnover of myofibrillar proteins but not of intermediate-filament proteins, *J. Cell Biol.*, 96, 684, 1983.

73. Joyce, N. C., Haire, M. F., and Palade, G. E., Contractile proteins in pericytes. I. Immunoperoxidase localization of tropomyosin, *J. Cell Biol.*, 100, 1379, 1985.
74. Zierler, R. E., Bandyk, D. F., Thiele, B. L., and Strandness, D. E., Carotid artery stenosis following endarterectomy, *Arch. Surg. (Chicago)*, 117, 1408, 1982.
75. Roederer, G. O., Langlois, Y. E., Lusiani, L., Jager, K. A., Primozich, J. F., Lawrence, R. J., Phillips, D. J., and Strandness, D. E., Natural history of carotid artery disease on the side contralateral to endarterectomy, *J. Vasc. Surg.*, 1, 62, 1984.
76. Roederer, G. O., Langlois, Y. E., Jager, K. A., Primozich, J. F., Lawrence, R. J., Beach, K. W., Phillips, D. J., and Strandness, D. E., The natural history of carotid arterial disease in asymptomatic patients with cervical bruits, *Stroke*, 15, 605, 1984.
77. Stoney, R. J. and String, S. T., Recurrent carotid stenosis, *Surgery*, 80, 705, 1976.
78. Hertzer, N. R., Martinez, B. D., and Beven, E. G., Recurrent stenosis after carotid endarterectomy, *Surg. Gynecol. Obstet.*, 149, 360, 1979.
79. Turnipseed, W. D., Berkoff, H. A., and Crummy, A., Postoperative occlusion after carotid endarterectomy, *Arch. Surg. (Chicago)*, 115, 573, 1980.
80. Kremen, J. E., Gee, W., Kaupp, H. A., and McDonald, K. M., Restenosis or occlusion after carotid endarterectomy, *Arch. Surg. (Chicago)*, 114, 608, 1979.
81. Imparato, A. M., Bracco, A., Kim, G. E., and Zeff, R., Intimal and neointimal fibrous proliferation causing failure of arterial reconstructions, *Surgery*, 72, 1007, 1972.
82. French, B. N. and Rewcastle, N. B., Sequential morphological changes at the site of carotid endarterectomy, *J. Neurosurg.*, 41, 745, 1974.
83. Cossman, D., Callow, A. D., Stein, A., and Matsumoto, G., Early restenosis after carotid endarterectomy, *Arch. Surg. (Chicago)*, 113, 275, 1978.
84. Clowes, A. W. and Clowes, M. M., The influence of chronic hypertension on injured and uninjured arteries in spontaneously hypertensive rats, *Lab. Invest.*, 43, 535, 1980.
85. Clowes, A. W., Reidy, M. A., and Clowes, M. M., Kinetics of cellular proliferation after arterial injury. I. Smooth muscle growth in the absence of endothelium, *Lab. Invest.*, 49, 327, 1983.
86. Reidy, M. A., Clowes, A. W., and Schwartz, S. M., Endothelial regeneration. V. Inhibition of endothelium in arteries of rat and rabbit, *Lab. Invest.*, 49, 569, 1983.
87. Orekhov, A. N., Kosykh, V. A., Repin, V. S., and Smirnov, V. N., Cell proliferation in normal and atherosclerotic human aorta. I. Flow cytofluorometric determination of cellular DNA content, *Lab. Invest.*, 48, 395, 1983.
88. Spagnoli, L. G., Villaschi, S., Neri, L., Palmieri, G., Taurino, M., Faraglia, V., and Fiorani, P., Autoradiographic studies of the smooth muscle cells in human arteries, *Paroi Arterielle/Arterial Wall*, 7, 107, 1981.
89. Villaschi, S. and Spagnoli, L. G., Autoradiographic and ultrastructural studies on the human fibroatheromatous plaque, *Atherosclerosis*, 48, 95, 1983.
90. Glenn, K. C. and Ross, R., Human monocyte-derived growth factor(s) for mesenchymal cells: activation of secretion by endotoxin and concanavalin A, *Cell*, 24, 603, 1981.
91. Martin, B. M., Gimbrone, M. A., Jr., Unanue, E. R., and Cotran, R. S., Stimulation of nonlymphoid mesenchymal cell proliferation by a macrophage-derived growth factor, *J. Immunol.*, 126, 1510, 1981.
92. Gabbiani, G., Kocher, O., Bloom, W. S., Vandekerckhove, J., and Weber, K., Actin expression in smooth muscle cells of rat aortic intimal thickening, human atheromatous plaque, and cultured rat aortic media, *J. Clin. Invest.*, 73, 148, 1984.
93. Wiener, J., Loud, A. V., Giacomelli, F., and Anversa, P., Morphometric analysis of hypertension-induced hypertrophy of rat thoracic aorta, *Am. J. Pathol.*, 88, 619, 1977.
94. Owens, G. K. and Schwartz, S. M., Vascular smooth muscle cell hypertrophy and hyperploidy in the Goldblatt hypertensive rat, *Circ. Res*, 53, 491, 1983.
95. Clowes, A. W. and Schwartz, S. M., Significance of quiescent smooth muscle migration in the injured rat carotid artery, *Circ. Res.*, 56, 139, 1985.
96. Reidy, M. A., personal communication, 1985.
97. Limas, C., Westrum, B., and Limas, C. J., The evolution of vascular changes in the spontaneously hypertensive rat, *Am. J. Pathol.*, 98, 357, 1980.
98. Chobanian, A. V., Prescott, M. F., and Haudenschild, C. C., Recent advances in molecular pathology. The effects of hypertension on the arterial wall, *Exp. Mol. Pathol.*, 41, 153, 1984
99. Schwartz, S. M. and Ross, R., Cellular proliferation in atherosclerosis and hypertension, *Prog. Cardiovasc. Dis.*, 26, 355, 1984.
100. Reidy, M. A. and Silver, M., Endothelial regeneration. VII. Lack of intimal proliferation after defined injury to rat aorta, *Am. J. Pathol.*, 118, 173, 1985.
101. Gajdusek, C. M. and Schwartz, S. M., Comparison of intracellular and extracellular mitogen activity, *J. Cell Physiol.*, 121, 316, 1984.

102. Barrett, T. B., Gajdusek, C. M., Schwartz, S. M., McDougall, J. K., and Benditt, E. P., Expression of the *sis* gene by endothelial cells in culture and *in vitro, Proc. Natl. Acad. Sci. U.S.A.*, 81, 6772, 1984.
103. Seifert, R. A., Schwartz, S. M., and Bowen-Pope, D. F., Developmentally regulated production of platelet-derived growth factor-like molecules, *Nature (London)*, 311, 669, 1984.
104. Folkow, B., Physiological aspects of primary hypertension, *Physiol. Rev.*, 62, 347, 1982.
105. Goldberg, I. D., Rosen, E. M., Shapiro, H. M., Zoller, L. C., Myrick, K., Levenson, S. E., and Christenson, L., Isolation and culture of a tetraploid subpopulation of smooth muscle cells from the normal rat aorta, *Science*, 226, 559, 1984.
106. Gordon, D., Mohai, L. G., and Schwartz, S. M., Induction of polyploidy in cultures of neonatal rat aortic smooth muscle cells, in preparation.
107. Salmons, A. and Sreter, F. A., Significance of impulse activity in the transformation of skeletal muscle type, *Nature (London)*, 263, 30, 1976.

Chapter 2

METHODS OF GROWING VASCULAR SMOOTH MUSCLE IN CULTURE

Julie H. Campbell and Gordon R. Campbell

TABLE OF CONTENTS

I. INTRODUCTION

The methodology, enzymes, culture medium, substrate, and chambers used for growing smooth muscle in culture vary with the type and purpose of the experiment, the age and species of the donor, and the personal preference of the investigators. In this chapter, the systems used by some of the established laboratories will be described. Most researchers use similar methods, modified slightly to suit their particular requirements.

II. CULTURE PROCEDURES

A. Enzyme-Dispersed Cells in Primary Culture

The methods developed in our laboratories will be described in detail. The enzymes used are collagenase (Worthington CLS III, approximately 105 units/mg) at 3 mg/mℓ in Medium 199 minus serum (M199-S) and elastase (Sigma E-0127, 60 units/mg) at 1 mg/mℓ in M199-S. Six hundred milligrams of collagenase is placed in 200 mℓ M199-S and dissolved by stirring over ice for 1 hr. In order to remove residual particulate material and sterilize the solution, it is filtered by negative pressure successively through 5 μm, 0.8 μm, 0.45 μm, and 0.2 μm Millipore® filters. It is then aliquoted into 20 × 10 mℓ sterile plastic tubes, frozen immediately at −20°C, and used within 1 month. 10 Milligram elastase is dissolved in 10 mℓ M199-S, and the pH is adjusted to 7 with sterile NaOH. The solution is sterilized by negative pressure through 0.2-μm Millipore® filters and used the same day.

1. Elastic Arteries (Pig, Rabbit, Rat Aortas)

Freshly dissected aortas from adult animals are placed in cold Hank's Balanced Salt Solution (HBSS) plus 5% fetal calf serum (FCS). The vessels are cleaned by removing the fat and connective tissue under a dissecting microscope, cut longitudinally; and, using watchmaker forceps, the intima plus the inner two thirds of the media is removed in 1-mm-wide horizontal strips from the adventitia. This procedure is carried out in a 90-mm glass petri dish containing HBSS-S and a scintered glass plate to prevent sliding of the tissue. The muscle strips are then transferred to collagenase in a 50-mℓ Erlenmeyer flask. The methodology can then proceed in either of two ways depending on the available time:

Method 1. The tissue plus collagenase is gently stirred on an orbital shaker for 1 hr at 37°C. The pH is regularly adjusted with 1 *M* NaOH. The collagenase is removed from the tissue with a Pasteur pipette and discarded, elastase added and the tissue pipetted up and down in a wide-bore pipette then stirred at 37°C for a further hour. Finally, more collagenase is added and the tissue stirred and regularly pipetted until all tissue has dispersed into single cells (about 1¹/₂ hr).

Method 2. If there is lack of time, an equal volume of M199-S is added to the tissue/collagenase and the flask placed in a 37°C incubator overnight. (The extra volume prevents an undesirable change in pH.) The following morning, fresh collagenase is added on top of the M199/collagenase/tissue, gently pipetted, and left for 30 min to 1 hr at 37°C with stirring until the tissue has completely dispersed into single cells.

2. Muscular Arteries (Rat Caudal Artery, Rabbit Ear Artery)

The vessels are cleaned of fat and connective tissue and placed in collagenase in a 60-mm petri dish. The vessels are stirred gently on an orbital shaker at 37°C. At regular 10-min intervals, the adventitia is very gently pipetted away from the underlying muscle. When the vessel becomes translucent, only the muscle layer is present. The vessels are placed in fresh collagenase, cut into 3-mm lengths, transferred to an Erlenmeyer flask with an equal volume of M199, and placed in a 37°C incubator overnight. The following morning the tissue suspension is centrifuged at 900 rpm for 5 min and the supernatant discarded. The tissue is resuspended in 0.125% trypsin-versene solution, gently stirred, and pipetted at 37°C until the cells are completely dissociated ($^1/_2$ to $^3/_4$ hr). FCS is added to inactivate the trypsin, and the suspension is again spun at 900 rpm for 5 min.

3. Arterioles (Rat Mesenteric Vessels)

Mesenteric vessels are embedded in a large amount of fat. To make the adipose tissue firm for ease of removal, the tissue is placed in HBSS on ice. It is then placed in collagenase at 37°C for 10 min, transferred back into cold HBSS, and the fat and connective tissue is stripped away. The cold HBSS is replaced frequently so that fat globules are removed. The mesenteric vein with its anastomosing branches are separated from the artery and arterioles and discarded. The mesenteric arterioles are now separated from the mesenteric artery and placed in 60-mm petri dishes with collagenase for 10 min so that the adventitia disperses away with gentle pipetting. The muscle layer is transferred to fresh collagenase, cut into small pieces, and dispersed into single cells the same as for muscular arteries.

4. Embryonic Blood Vessels (Chick Aorta)

Embryonic and neonatal blood vessels disperse very easily and require only a 30-min incubation in collagenase (discarded), followed by successive 10-min incubations in 0.1% trypsin (1:250; Nutritional Biochemicals) in Ca^{2+} and Mg^{2+}-free HBSS.

5. Plating of Dispersed Cells

The enzyme-cell suspensions are transferred to 10-mℓ centrifuge tubes containing 1 mℓ of FCS and spun at 900 rpm for 5 min. A drop of the supernatant is placed on a microscope slide to check that all cells have pelleted. If some cells remain in suspension, the supernatant is transferred to fresh tubes and respun at 1200 rpm. The cell pellets are gently resuspended in nutrient medium, a cell count performed, then the cell concentration adjusted to the required amount and the cells plated out.

References 1 to 13 provide variations of all of the above methods.

B. Reaggregrates of Enzyme-Dispersed Cells

The methodology described is from Sperelakis' laboratory.[3-5]

The blood vessels are dispersed into single cells, pelleted, and resuspended in nutrient medium to a concentration of about 10^6 cells per milliliter. Spherical reaggregates of the separated cells are obtained by two methods:

Method 1 — One milliliter of cell suspension is placed in a well-chamber made from a glass ring attached to boiled cellophane dialysis tubing (2 cm diameter) by high-vacuum silicone grease, enclosed within a Carrel flask. After 2 mℓ of culture medium is placed around the glass rings in the Carrel flask, the culture is incubated at 37°C. The cells do not adhere to the dialysis tubing, and reaggregates of dispersed cells begin to form within 36 hr. After 3 to 5 days, the reaggregates (50 to 200 μm diameter) become round, translucent, and firmly packed.

Method 2 — In this method, which is based on that of Moscona,[14] 2 to 3 mℓ of cell suspension is placed in 25-mℓ Erlenmeyer flasks, sealed with rubber stoppers, and covered with parafilm. The flasks are rotated in a gyratory shaker bath at 37°C for 48 hr at a speed of 70 rpm.

C. Free-Floating Explants

The methodology is from Jarmolych and colleagues[15-17] (see also Reference 18).

Medial tissue is cut into segments (1 × 1 mm to 1 × 5 mm) and placed in 30-mℓ plastic culture bottles containing 5 mℓ of nutrient medium. The explants do not attach to the substrate, and there is no cell outgrowth beyond the periphery of the explant itself.

D. Substrate-Attached Explants

The methodology is from Wissler and colleagues.[19,20]

Round explants, 2 mm in diameter, are punched out from cleaned medial tissue with a sterile skin biopsy punch. These explants are placed in 30-mℓ plastic flasks, usually one in each flask, and incubated in 2.5 mℓ medium plus serum at 37°C. To facilitate adherence of the explants to the substrate, the flasks are turned upside down for 2 hr each day. Initial outgrowth occurs within 1 to 2 weeks.

E. Subcultured Cells

The methodology is from Ross.[21,22]

Intimal-medial strips are placed in a petri dish containing nutrient medium and cut into 1-mm^3 pieces using a scissoring action with two Number 11 Bard-Parker surgical blades. Explants are cut cleanly without tearing as cells do not appear to migrate readily from torn edges of tissue. The explants are washed several times in nutrient medium and dispensed into pregassed, 75-cm^2 plastic flasks with 3 mℓ of medium containing 100 to 150 explants per flask. The flasks are swirled to disperse the explants evenly. When the explants have attached to the bottom of the flask (about 1 week), a greater volume of medium is added. Alternatively, the flask can be maintained upright for 2 to 3 hr to encourage attachment of explants.[23] A coverslip[24] or dialysis membrane[25] is placed over the explants to hold them in place, or each explant is immobilized into the plastic surface of the flask with a platinum-Teflon®-steel wire positioner.[26] Outgrowth from the explants reaches confluency in 2 to 3 weeks. They are then washed in Tris-versene buffer (0.54 m*M* versene and 0.02 *M* Tris-HCl pH 7.4 at 37°C) and detached from the substrate with 0.05% trypsin (Difco® 1:250 trypsin) in Tris-versene buffer by incubation for 10 min at 37°C. The trypsin is inactivated with medium containing 10% FCS, and the suspension is centrifuged at 1000 rpm for 7 min. The cell pellet is resuspended in medium plus serum and dispensed into culture dishes at a known concentration. When confluency is reached, the cells are subcultured, using the trypsinization procedure described above.

III. CULTURE MEDIUM

A. Composition

The most commonly used medium for growth of vascular smooth muscle in culture is the Dulbecco-Vogt modification of Eagle's minimal essential medium[27] plus 5 or 10% serum. Frequent additions are 2 to 4 m*M* glutamine, 0.28 *M* bicarbonate, and 100 m*M* nonessential amino acids;[28] sometimes, 1 m*M* sodium pyruvate is also added.[21] Eagle's minimal essential medium, either with[23] or without[10,19] supplements, is also popular. Several laboratories, including our own, use Medium 199 with Hank's salts,[1,15-18] while in other laboratories Ham's F10[29] and Ham's F12[26,30] media are used. Antibiotics are used universally, generally 100 Iu/mℓ penicillin plus 100 μg/mℓ streptomycin. Some laboratories use 50 Iu/mℓ penicillin plus 50 μg/mℓ streptomycin plus 100 μg/mℓ kanamycin, while others use 50 μg/mℓ gentamycin alone. In our laboratories, 100 Iu/mℓ penicillin alone is used.

B. Serum

For convenience and good growth characteristics, most laboratories use either 5 or

10% FCS (whole blood serum) in nutrient medium for vascular smooth muscle cultures. However, because of the escalating cost of this product, many laboratories are making their own serums. This is not difficult and has the advantage that serum from the same species from which the cells are derived can be used. Additionally, as pointed out by Ross et al.[31] (see also Chapter 6, this volume), vascular smooth muscle cells (SMC) in vivo come into contact only with whole blood serum when platelet aggregation and release has occurred in a vessel with an injured endothelium. Under normal circumstances, SMC are bathed by a filtrate of plasma. For vascular smooth muscle cultures, Ross has pioneered the use of homologous plasma-derived serum, which does not contain platelet-releaseate products, and in which the cells remain quiescent as in vivo.

The methodology for preparation of whole blood serum and plasma-derived serum comes from Ross and Kariya.[21]

1. Whole Blood Serum for Stock Cultures

Blood from several animals is pooled and allowed to clot for at least 24 hr at 4°C. The resulting serum is separated from the clots, pooled, and centrifuged at 22,000 × g for 15 min at 4°C. The centrifuged serum is filtered through a 0.22-μm Millipore filter using positive pressure and heat inactivated at 56°C for 30 min; it is then stored at −70°C.

2. Whole Blood Serum

Using plastic or siliconized glass syringes, the blood is drawn from at least four animals into 3.8% sodium citrate, one volume citrate to ten volumes blood. Each donor's blood is separately calcified with 1.0 *M* $CaCl_2$ to a final concentration of 14 μ mol/mℓ allowed to clot at 37°C for 2 hr, and then centrifuged at 1085 × g for 15 minutes at 4°C. The serums are respun at 22,000 × g for 30 min at 4°C, pooled, and dialyzed (Dialya Por, 6000 to 8000 mol wt cutoff) against Ringers at 4°C for 24 hr. It is then heat inactivated at 56°C for 30 min, filtered through 0.22-μm Millipore®, and stored at −70°C.

3. Plasma-Derived Serum

The blood is drawn into prechilled (4°C) citrated plastic or siliconized glass syringes. The blood from each donor is spun separately in plastic or siliconized glass tubes at 1085 × g for 15 min at 4°C and the resulting plasma respun at 22,000 × g for 30 minutes at 4°C. The platelet-poor plasma is pooled and kept on ice for 3 hr; 1.0 *M* $CaCl_2$ to a final concentration of 20 μ mol/mℓ and one volume of Ringers for six volumes plasma are added; then the plasma is incubated at 37°C for 2 hr. The plasma is spun at 22,000 × g for 30 min to remove the fibrin clot and dialyzed against 0.1 *M* Tris pH 7.4 overnight. The plasma-derived serum is applied to a CM-Sephadex C-50 column and eluted with 0.1 *M* Tris pH 7.4. The eluate is concentrated back to its original volume by ultrafiltration using an Amicon Diaflo PM-10 membrane, dialyzed for 24 hr against Ringers at 4°C, heat inactivated, and stored at −70°C.

IV. CULTURE CHAMBERS AND SUBSTRATES

Vascular SMC are generally grown in plastic culture dishes or flasks; however, they also attach well to very clean glass coverslips placed either in plastic dishes or in specialized chambers, such as the Rose chamber. They grow well on collagen membranes,[32] elastic membranes,[33] smooth silicone rubber substrata,[34] polyacrylonitrile (orlon) fabrics, dacron velour, and Parylene-C coated polypropylene microfabric,[35] and microcarrier beads (see Chapter 3).

ACKNOWLEDGMENTS

The expert assistance of Janet Rogers, Lucy Popadynec, Liana Butera, Bridget Southwell, and Mary Jane Black is gratefully acknowledged.

REFERENCES

1. Chamley, J. H., Campbell, G. R., McConnell, J. D., and Gröschel-Stewart, U., Comparison of vascular smooth muscle cells from adult human, monkey and rabbit in primary culture and in subculture, *Cell Tissue Res.*, 177, 503, 1977.
2. Chamley-Campbell, J. H., Campbell, G. R., and Ross, R., The smooth muscle cell in culture, *Physiol. Rev.*, 59, 1, 1979.
3. Harder, D. R. and Sperelakis, N., Action potential generation in reaggregates of rat aortic smooth muscle cells in primary culture, *Blood Vessels*, 16, 186, 1979.
4. Ives, H. E., Schultz, G. S., Galardy, R. E., and Jamieson, J. D., Preparation of functional smooth muscle cells from the rabbit aorta, *J. Exp. Med.*, 148, 1400, 1978.
5. McLean, M. J. and Sperelakis, N., Electrophysiological recordings from spontaneously contracting reaggregates of cultured vascular smooth muscle cells from chick embryos, *Exp. Cell Res.*, 104, 309, 1977.
6. Martin, G. M., Ogburn, C. E., and Wight, T. N., Comparative rates of decline in the primary cloning efficiencies of smooth muscle cells from the aging thoracic aorta of two murine species of contrasting maximum life span potentials, *Am. J. Pathol.*, 110, 236, 1983.
7. Hermsmeyer, K., De Cino, P., and White, R., Spontaneous contractions of dispersed vascular muscle in cell culture, *In Vitro*, 12, 628, 1976.
8. Gimbrone, M. A. and Cotran, R. S., Human vascular smooth muscle in culture. Growth and ultrastructure, *Lab. Invest.*, 33, 16, 1975.
9. Nilsson, J., Ksiazek, T., Heldin, C.-H., and Thyberg, J., Demonstration of stimulatory effects of platelet-derived grov·th factor on cultivated rat arterial smooth muscle cells. Differences between cells from young and adult animals, *Exp. Cell Res.*, 145, 231, 1983.
10. May, J. F., Paule, W. J., Rounds, D. E., Blankenhorn, D. H., and Zemplenyi, T., The induction of atherosclerotic plaque-like mounds in cultures of aortic smooth muscle cells, *Virchows Arch. B*, 18, 205, 1975.
11. Garay, R. P., Moura, A. M., Osborne-Pellegrin, M. J., Papadimitriou, A., and Worcel, M., Identification of different sodium compartments from smooth muscle cells, fibroblasts and endothelial cells, in arteries and tissue culture, *J. Physiol. (London)*, 287, 213, 1979.
12. Yau-Young, A. O., Shio, H., and Fowler, S., Growth, biochemistry, and morphology of isolated rabbit aortic smooth muscle cells maintained in the presence or absence of serum, *J. Cell. Physiol.*, 108, 461, 1981.
13. Martin, W. and Gordon, J. L., Spontaneous and agonist-induced ^{86}Rb efflux from rabbit aortic smooth muscle cells in culture: a comparison with fresh tissue, *J. Cell. Physiol.*, 115, 53, 1983.
14. Moscona, A., Rotation-mediated histogenic aggregation of dissociated cells. A quantifiable approach to cell interactions in vitro, *Exp. Cell Res.*, 22, 455, 1961.
15. Jarmolych, J., Daoud, A. S., Landau, J., Fritz, K. E., and McElvene, E., Aortic media explants. Cell proliferation and production of mucopolysaccharides, collagen, and elastic tissue, *Exp. Mol. Pathol.*, 9, 171, 1968.
16. Daoud, A. S., Fritz, K. E., and Jarmolych, J., Increased DNA synthesis in aortic explants from swine fed a high-cholesterol diet, *Exp. Mol. Pathol.*, 13, 377, 1970.
17. Fritz, K. E., Jarmolych, J., and Daoud, A. S., Association of DNA synthesis and apparent dedifferentiation of aortic smooth muscle cells *in vitro*, *Exp. Mol. Pathol.*, 12, 354, 1970.
18. Rossi, G. L., Alroy, J., and Röthenmund, S., Morphological studies of cultured swine aorta media explants, *Virchows Arch. B*, 12, 133, 1973.
19. Fisher-Dzoga, K., Jones, R. M., Vesselinovitch, D., and Wissler, R. W., Ultrastructural and immunohistochemical studies of primary cultures of aortic media cells, *Exp. Mol. Pathol.*, 18, 162, 1973.
20. Fisher-Dzoga, K., Chen, R., and Wissler, R. W., Effects of serum lipoproteins on the morphology, growth, and metabolism of arterial smooth muscle cells, in *Arterial Mesenchyme and Arteriosclerosis*, Wagner, W. D. and Clarkson, T. B., Eds., Plenum Press, New York, 1974, 299.

21. Ross, R. and Kariya, B., Morphogenesis of vascular smooth muscle in atherosclerosis and cell culture, in *Handbook of Physiology,* Section 2: The Cardiovascular System, *Vascular Smooth Muscle,* Vol. 2, Bohr, D., Somlyo, A. P., and Sparks, H. V., Jr., Eds., American Physiological Society, Washington, D.C., 1980, 69.
22. Ross, R., The smooth muscle cell. II. Growth of smooth muscle in culture and formation of elastic fibers, *J. Cell Biol.,* 50, 172, 1971.
23. St. Clair, R. W., Smith, B. P., and Wood, L. L., Stimulation of cholesterol esterification in Rhesus monkey arterial smooth muscle cells, *Circ. Res.,* 40, 166, 1977.
24. Brown, B. G., Mahley, R., and Assman, G., Swine aortic smooth muscle in tissue culture. Some effects of purified swine lipoproteins on cell growth and morphology, *Circ. Res.,* 39, 415, 1976.
25. Chamley, J. H., Goller, I., and Burnstock, G., Selective growth of sympathetic nerve fibers to explants of normally densely innervated autonomic effector organs in tissue culture, *Dev. Biol.,* 31, 362, 1973.
26. Björkerud, S. and Ekroth, R., The growth of human atherosclerotic and non-atherosclerotic aortic intima and media in vitro, *Artery,* 8, 329, 1980.
27. Vogt, M. and Dulbecco, R., Steps in the neoplastic transformation of hamster embryo cells by polyoma virus, *Proc. Natl. Acad. Sci. U.S.A.,* 49, 171, 1963.
28. Pietilä, K. and Nikkari, T., Effect of phase of growth and hyperlipidemic serum on the synthesis of collagen in rabbit aortic smooth muscle cells in culture, *Med. Biol.,* 56, 111, 1978.
29. Larrue, J., Daret, D., Demond-Henri, J., Allières, C., and Bricaud, H., Prostacyclin synthesis by proliferative aortic smooth muscle cells. A kinetic in vivo and in vitro study, *Atherosclerosis,* 50, 63, 1984.
30. Thyberg, J., Nilsson, J., Palmberg, L., and Sjölund, M., Adult human arterial smooth muscle cells in primary culture. Modulation from contractile to synthetic phenotype, *Cell Tissue Res.,* 239, 69, 1985.
31. Ross, R., Nist, C., Kariya, B., Rivest, M.-J., Raines, E., and Callis, J., Physiological quiescence in plasma-derived serum: influence of platelet-derived growth factor on cell growth in culture, *J. Cell. Physiol.,* 97 (Suppl.1), 497, 1978.
32. Csonka, E., Szemenyei, K., Miskulin, M., and Robert, A. M., Morphological examination of aortic endothelial and smooth muscle cells grown *in vitro* on collagen membranes, *Artery,* 8, 253, 1980.
33. Leung, D. Y. M., Glagov, S., and Mathews, M. B., Cyclic stretching stimulates synthesis of matrix components by arterial smooth muscle cells *in vitro, Science,* 191, 475, 1976.
34. Buck, R. C., Behaviour of vascular smooth muscle cells during repeated stretching of the substratum *in vitro, Atherosclerosis,* 46, 217, 1983.
35. Eskin, S. G., Armeniades, C. D., Lie, J. T., Trevino, L., and Kennedy, J. H., Growth of cultured calf aortic smooth muscle cells on cardiovascular prosthetic materials, *J. Biomed. Mater. Res.,* 10, 113, 1976.

Chapter 3

MICROCARRIER CULTURE OF VASCULAR SMOOTH MUSCLE CELLS

Susan E. O'Connor and Peter F. Davies

TABLE OF CONTENTS

I. INTRODUCTION

The in vitro culture of vascular smooth muscle cells (SMC) is an integral component of cardiovascular research in relation to both normal vascular physiology and the pathogenesis of atherosclerosis. Any procedures which offer new and potentially valuable methods for propagating vascular cells are therefore of significant interest. Microcarrier systems, in addition to providing scaled-up culture of vascular cells, enable a variety of innovative approaches to vascular cell research. In this chapter we present straightforward techniques for the propagation on microcarriers of anchorage-dependent cells in general and vascular SMC in particular, using methods in routine use in this laboratory.

II. CONCEPT

Anchorage-dependent cells die quickly without attachment to a substratum and therefore cannot be maintained in suspension culture. In an attempt to overcome this limitation, the use of small beads to propagate anchorage-dependent cells in suspension culture was first introduced by van Wezel[1] in 1967. Significant advances in the development of microcarrier technology, particularly since the mid-1970s, have resulted in the culture of many different kinds of cells on beads in suspension. Development was stimulated primarily by a need to scale up cultures of anchorage-dependent animal cell lines for the efficient production of biological products. Examples include the production of vaccines,[2-5] interferon,[6] and other commercially valuable cell products.[7-11] The advantages of scale-up have also been utilized to obtain large numbers of cells synchronized in mitosis[12] and high yields of purified membrane fractions.[13]

Very recently, novel manipulations of relatively small quantities of cells have been used for research purposes. Cardiovascular research, in particular, has made use of a number of microcarrier types to address questions relevant to vascular cell function that were previously difficult to investigate. For example, Ryan et al.[14] have used a microcarrier system to monitor cell-surface events in pulmonary artery endothelium. Previous studies were hindered by the need for proteolytic enzyme digestion during isolation and subculture procedures. Furthermore, the ease of transfer of anchorage-dependent cells has been effectively utilized — first, to prepare microcarrier beds in glass columns for biochemical studies in endothelial cells using flow-through systems[15,16] and, second, in studies of vascular cell interactions[17,18,34,35] (see also Davies, Volume II, Chapter 15).

III. MICROCARRIERS

A. General Characteristics

The earliest microcarriers developed from DEAE Sephadex[1] possessed high positive charges distributed throughout a porous matrix. The adsorption of nutrients and cofactors from the culture medium onto a large internal surface area and the restriction by high electrostatic interactions of normal cell movements led to poor plating efficiencies and low replication rates. It has since been shown that the lowest possible positive charge or low negative charge is optimal for cell attachment and spreading.[19] When the charge is limited to the surface of the microcarrier, adsorption of serum proteins by the large internal surface area of a porous bead is markedly reduced.[20] The introduction of solid beads minimized these problems and also reduced the volume of culture medium required to bathe the cells by eliminating diffusion of medium into the bead. The smooth surface of solid microcarriers also promotes efficient cell attachment and spreading.

The composition of the bead matrix determines the ease of preparation of microcarriers prior to cell culture. Hydration procedures involving multiple washings are inconvenient features of some microcarriers. If prepared swollen beads are stored, bactericidal agents must be removed before their use with cultured cells. The most convenient forms of microcarriers are solid matrices supplied in a sterilized, dry form.

Because microcarriers are best suited for suspension culture, the specific density of the beads must allow them to be easily mixed at low stirring speeds. Most fall in the density range of 1.03 to 1.045 g/mℓ. Those at the higher end of the range rapidly settle upon cessation of stirring; although this can be a useful feature for the separation of medium and beads, such microcarriers often require higher stirring speeds. The density of the microcarrier changes with the addition of cells and their subsequent proliferation. With increasing cell numbers, it is sometimes necessary to increase the stirring speed to maintain the microcarriers in suspension.

For visualizing cells on microcarriers either by phase-contrast microscopy of living cells or after staining, a transparent microcarrier is useful. Under microscopic observation, two planes of focus are possible, giving visual access to both the "front" and "rear" of the microcarrier and enabling a simple estimate to be made of the extent of colonization.

B. Commercially Available Microcarriers

The different types of commercially available microcarriers are listed in Table 1 along with their major properties. While all of the microcarriers listed will support the attachment and growth of anchorage-dependent cells, some are more suited for use with either epithelial or fibroblast-like cells. Biocarrier, Cytodex 1, Superbeads, and Cytodex 2 are reported to be more useful with cells having a fibroblast-like morphology, while Cytodex 3, Biosilon, and Cytosphere are suggested for cells with epithelial morphology.[21,22] The choice of microcarrier, however, is best made with the type of experiment in mind since all will, with appropriate handling, provide viable cultures. For example, is an intrabead porous volume important for the experiments? Are morphological studies (including scanning and transmission electron microscopy) important, and will bead shrinkage compromise their interpretation?

For vascular cell culture, our laboratory has had experience primarily with Biosilon (Nunc, Denmark; Vangard International, Inc., Neptune, N.J.). This choice arose simply because these microcarriers, which we use extensively for studies of cellular biochemistry, have the closest composition to tissue culture plastic and, in the system described below, reproducibly propagate vascular endothelial and SMC, skin fibroblasts, and 3T3 cells with identical characteristics (as far as we can determine) to the same cells grown in dishes and flasks. They also maintain suspension cultures of adherent human blood monocytes and mouse peritoneal macrophages for short periods of time. Properties which render this microcarrier particularly suitable for biochemical experiments with cultured cells are

1. Nonporous solid polystyrene (Figure 1) of similar composition to standard tissue culture dishes and flasks (the nonporous nature eliminates high adsorption of radiolabeled tracers and metabolic products; cells are readily removed from the plastic surface by trypsin.EDTA for cell counts and for the measurement of cell-to-cell bead ratios)
2. Transparent, nonshrinking upon air-drying, and extractable into organic solvents thereby allowing cells to be fixed and stained for observation and processed for electron microscopy without shrinkage

Table 1
COMMERCIAL MICROCARRIERS

Name	Manufacturer	Type[a]
Biosilon	Nunc, Denmark	Polystyrene, NP
Biocarriers	Biorad, U.S.	Polyacrylamide, P
Cytodex 1, 2, 3	Pharmacia, Sweden	Dextran, P
Cytospheres	Lux, U.S.	Polystyrene, NP
DE-52, DE-53	Whatman, U.K.	Microgranular cellulose, P
Gelibeads	KC Biological, U.S.	Cross-linked gelatin, P
Superbeads	Flow Labs, U.S.	Dextran, P

[a] P, porous matrix; NP, non-porous matrix.

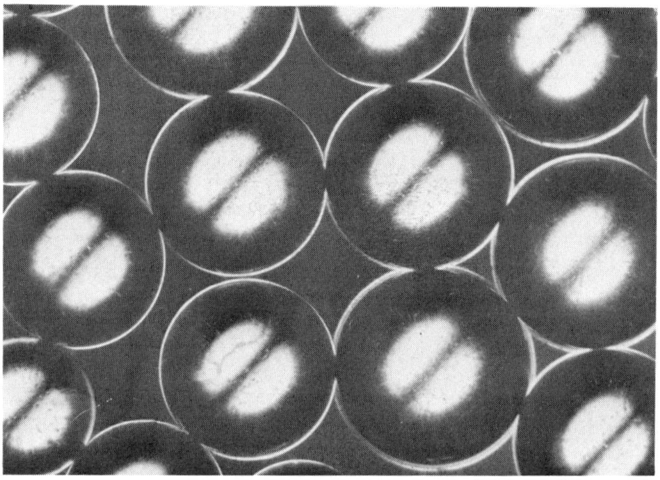

FIGURE 1. Solid polystyrene microcarriers (Biosilon; Nunc, Denmark) suitable for SMC culture. Average diameter, 208 μm. Photographed by dark-field microscopy. (Magnification × 125.)

A reported disadvantage of these microcarriers is the sensitivity of anchored cells to shear forces developed during stirring.[20] This may lead to cell loss and cell injury resulting from collisions of the rigid spheres. In practice, however, we have found that an appropriate choice of stirring system eliminates these failings.

IV. STIRRER SYSTEMS

Microcarriers are routinely maintained in suspension culture (rather than in perfused columns or stationary cultures) because conditions for efficient nutrient and metabolite diffusion are met by the constant movement of the stirred medium. The relatively large numbers of cells grown require uniform exposure to fresh medium throughout the culture flask. For laboratory-scale cultures which can be operated in standard incubators, the primary method for maintaining a suspended microcarrier culture is the stirrer system.

The requirements of a stirrer system for this purpose are numerous. First and foremost, it must keep all of the microcarriers in suspension at stirring speeds consistent with good cell growth and replication. The microcarriers should be in a homogeneous solution with no pockets of beads left stagnant or trapped in a limited area. This ensures that the cells are all maintained in the same environment experiencing similar low-shear forces. The stirrer flask itself must allow for gas exchange and be able to maintain a suitable pH. Accessible sample ports and ease of cleaning and sterilization are also important.

Available systems range from simple paddle stirrers to sophisticated designs which impose more gentle conditions of culture. The earliest microcarrier cultures were grown in spinner flasks designed for use with transformed cells normally grown in suspension. Hirtenstein et al.,[23] however, have observed that optimal conditions for the suspension of single cells are probably different from those required for microcarriers. For example, simple paddle stirrers often have too high a stirring speed for microcarriers and the presence of an immersed spinner-bearing that tends to crush circulating beads. For these reasons a modified spinner vessel was introduced (Bellco Glass, Inc.). These flasks made use of a paddle attached to the impeller, which allowed the microcarriers to be suspended at a lower stirring speed. This change resulted in an increase in the yield of several cell types. Unfortunately, the paddles produced uneven flow patterns that created different local environments within the flask.[23] To address this problem, Techne Ltd. (Cambridge, U.K.) introduced a rod-stirred culture system with an indented base. As Figure 2 indicates, the rod is attached by a rubber gasket to a screw cap. The bulb at the end of the rod contains a magnet that is drawn around by another magnet placed in the stand on which the flask is mounted. The indented base creates a trough which conforms to the shape of the bulb. This combination of rod and trough forces the medium to circulate in a manner which uniformly suspends the microcarriers. Figure 3 illustrates the pattern of flow of an ink drop placed on the surface of a stirred 0.4% polyacrylamide solution. The motion of the rod and bulb forces the liquid to flow in a peripheral downward helical path and then up to the surface in a spiral with ever-increasing radius.[24] Horizontal and vertical circulation creates a constantly moving solution with no static area beneath the stirrer, as is common with flat-bottomed flasks. The Techne system includes a "softstart" facility that gently pushes the microcarriers into suspension, avoiding excessive turbulence. The entire Techne MCS-104 system is engineered for use in an incubator; the magnetic stirrer produces very little heat or noise and accommodates four flasks. We have found this particular system to be well-suited to the growth and maintenance of vascular SMC and endothelial cells.[18]

Our microcarrier cultures are routinely grown in 1 ℓ Techne stirrer flasks that have a working volume of 500 mℓ. Before use, the interior of each flask is treated with a silicon solution (Prosil®-28, SCM Corporation, Gainesville, Fla.; 1% in distilled water) and rinsed with distilled water. This procedure is repeated when necessary to prevent cell attachment to the walls of the stirrer flask.

V. ROLLER BOTTLES

Roller bottles may also be used with microcarrier cultures, although cell yields are less than those reported for spinner and rod-stirred cultures,[21] and special incubators are usually required. The microcarriers can be used to increase the surface area of the roller bottle by the simple addition of empty beads, along with the medium or initial cell inoculum. The microcarriers will adhere to the walls of the bottle, and the rolling speed is maintained optimal for each individual cell type. Siliconized roller bottles may be used for traditional suspension cultures; however, this requires that the bottles be completely filled with medium.

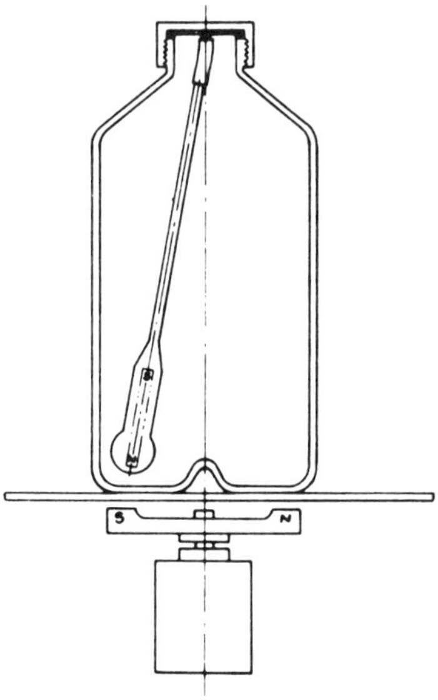

FIGURE 2. Outline of Techne MCS-104 microcarrier stirring system. A bulbous rod is attached to the inside of the cap by a length of silicon rubber. A magnet in the bulb induces pendulous rotation by attraction for another magnet in the base turned by an electric motor. (From De Bruyne, N. A. and Morgan, B. J., *Am. Lab.*, June 1981. With permission.)

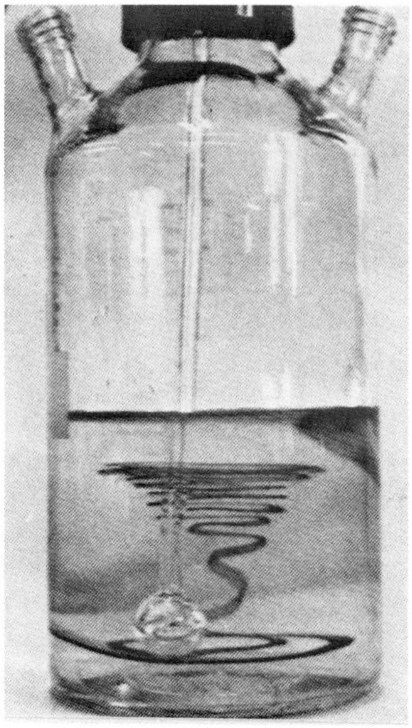

FIGURE 3. Helical path followed by an ink drop placed on the surface of a stirred 0.4% polyacrylamide solution in the MCS-104 stirrer. A combination of primary horizontal rotation and secondary motion at the start of the upward climb results in outward movement towards the surface. (From De Bruyne, N. A. and Morgan, B. J., *Am. Lab.*, June 1981. With permission.)

VI. SHORT-TERM STATIC CULTURES

For pilot studies, it is often useful to initiate small microcarrier cultures in static conditions. Microcarrier cultures of this type are plated and grown in tissue culture dishes that have been treated with an inert polymer, hydroxyethyl methacrylate (Hydron Laboratories, New Brunswick, N.J.; 1.2% solution in ethanol) to prevent the cells from attaching to the bottom of the dish.[16] Bacteriological plastic is also suitable to prevent adhesion to petri dish surfaces. Under static conditions, many more cells than usual grow in the petri dish. Therefore, the medium must be changed frequently to maintain the proper pH and adequate diffusion of nutrients and metabolites.

VII. CULTURES OF VASCULAR SMOOTH MUSCLE CELLS ON MICROCARRIERS

A. Smooth Muscle Cell Propagation

SMC obtained by nonenzymatic explant culture[25] and primary culture through enzymatic digestion[26] are suitable for use on microcarriers. The techniques described below are standard in our laboratory and are appropriate for enzymatically digested primary cells in suspension or cells which have been removed from standard tissue culture dishes by trypsinization.

All procedures are carried out in a laminar flow hood. Microcarriers (Biosilon), which are supplied dry and sterile, are simply transferred to a Techne stirrer flask (capacity 1ℓ) containing 250 mℓ of culture medium with 10% serum several hours before use. The serum concentration should be increased to 20% for primary cells or for other cells with low plating efficiency. Presoaking in a growth medium enhances attachment by coating the beads with plasma cold-insoluble globulins. These amphoteric bridging proteins link the negatively charged bead surface to the net negatively charged cell surface. In the absence of linking protein (either from serum or pretreatment of the beads with purified fibronectin), plating efficiency is very poor.[16]

A suspension of SMC (obtained through the methods described above) is added to the stirrer flask containing the soaked microcarriers, and the cell/bead suspension may be left static or, if automation is available, stirred intermittently (3 min on, 2 min off) for up to 24 hr. The flask is then filled to a final volume of 500 mℓ and left stirring. Hirtenstein et al.[23] have emphasized the importance of maintaining a head space above the culture medium that is at least as great in volume as the culture medium below it. Our experience has shown that a speed of 40 to 50 rpm in the Techne stirrer system is optimal to maintain all the beads in suspension while minimizing shear effects. Recently, a stirring speed of 18 rpm has been reported optimal for endothelial cells grown on Cytodex 3 microcarriers.[27]

The microcarriers should be inoculated with between 2×10^6 and 7×10^6 cells per gram of beads to achieve confluence within 5 to 10 days. This initial plating density of between 5 and 20 cells per bead will yield a confluent culture of about 150 to 200 cells per bead. An example of a growth curve for aortic SMC is shown in Figure 4. The number varies with the size distribution of the beads, the age of the culture, and the extent of bead clumping (see later). One gram of dry Biosilon microcarriers contains approximately 3.7×10^5 beads, giving a total surface area of 490 cm^2.

SMC are readily maintained in a growing state in Dulbecco's Modified Eagle's Medium containing 10% calf serum (M.A. Bioproducts, Watersville, Md.), 2 μmol/mℓ glutamine and 100 units each of penicillin and streptomycin. If a quiescent (nongrowing) culture of SMC is desired, the microcarriers are washed with Hank's Balanced Salt Solution (HBSS) to remove most serum growth factors and kept in 5% Plasma-derived serum prepared according to the methods of Vogel et al.[28]

B. Stain for Light Microscopy

Limited observation of the microcarriers under phase microscopy is possible. The perimeter of the sphere is visible, and the cells are seen as "bumps" along this edge (Figure 5). The orientation of the edge is random; therefore, some impression of the extent of colonization of the beads is possible. To visualize the cells in more detail, it is necessary to fix and stain them. A sample of microcarriers is rinsed several times with HBSS to remove the culture medium. The microcarriers are then fixed for 2 to 3 min in 95% ethanol and stained for 3 to 4 min in hematoxylin.* These procedures can be readily carried out in a conical centrifuge tube; the beads settle rapidly allowing removal of unwanted supernatant. After rinsing off the excess hematoxylin with distilled water, the stained beads are placed on a microscope slide and observed under

* With the aid of heat, dissolve 50 g of hematoxylin crystals in 50 mℓ of 100% alcohol and 100 g of alum in 1000 mℓ distilled water. Remove from heat and mix the two solutions together. Bring solution to a rapid boil (less than 1 min with stirring) and remove from heat. Add 2.5 g mercuric oxide (red) slowly and reheat to a simmer until it becomes a dark purple. Immediately plunge beaker in ice water. Stain is ready for use when cool. Addition of 3 mℓ glacial acetic acid per 100 mℓ of solution increases the precision of nuclear stain.[25]

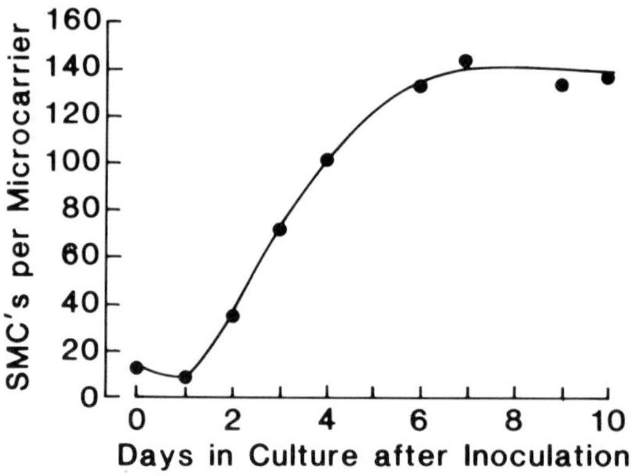

FIGURE 4. Growth curve of bovine aortic SMC isolated by the explant technique (25) and plated on Biosilon microcarriers. Initial plating density was ten cells per bead. Confluence (120 cells per bead) occurred after 6 days of suspension culture.

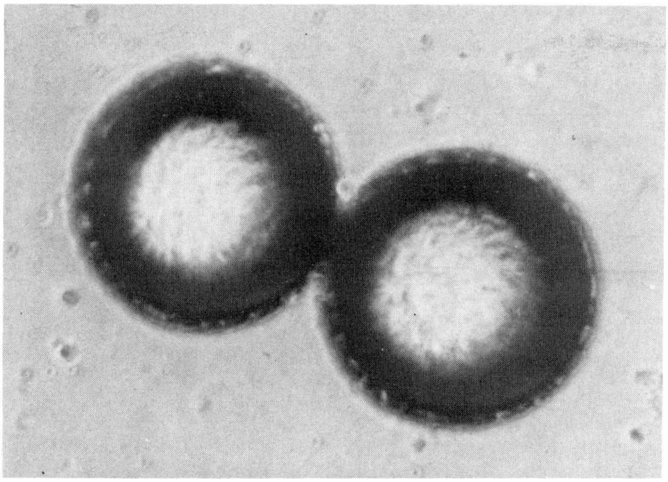

FIGURE 5. Phase-contrast microscopic appearance of vascular SMC at subconfluent density (90 cells per Biosilon bead). Although cells occupying the front and rear faces of each microcarrier are not clearly seen, an estimate of the extent of colonization can be made from the bead periphery. (Magnification unstained × 175.)

10X power with a standard light microscope. If a permanent record is desired, the beads can be covered with a thin layer of mounting medium (e.g., per mount).

Measurement of cell numbers following trypsinization is similar to that used with standard tissue culture flasks because the microcarriers are also made of polystyrene. A sample of microcarriers is placed in a conical tube, the microcarrier-bound cells are removed with trypsin: EDTA, gently vortexed, and the beads are allowed to settle to the bottom of the tube. The released cells remain in suspension much longer than the beads. The cell number in the supernatant is then determined using a hemocytometer or Coulter Counter (Coulter Electronics, Hialeah, Fla.). The bead number in the same sample can be determined by counting a diluted aliquot of resuspended beads (vortex) in the light microscope. The cell-to-bead ratio can then be calculated.

Total cell protein is determined by dissolving microcarrier-bound cells in 0.1% sodium dodecyl sulfate (30 min at room temperature) with occasional vortexing. The insoluble beads settle, and the supernatant is assayed by the Lowry procedure.[29]

C. Descriptive Morphology of Smooth Muscle Cells on Microcarriers

At a plating density of 5 to 20 cells per bead, individual SMC assume a fusiform morphology with random orientation (Figures 6a and 6b). As the beads become more heavily colonized (Figures 6c and 6d), the fusiform shape becomes less apparent. A characteristic feature of vascular SMC in culture is the formation of a "hill and valley" conformation after reaching confluence. We have never observed multilayers of SMC on single microcarriers although multilayers will form on occasion between beads if "bridging" is extensive. One may speculate that the inhibition of multilayer formation is associated with significant shear forces generated in stirred suspension. Once a confluent density is achieved, it is common for two or more microcarriers to become "bridged" by migrating SMC. These cells have an elongated morphology and are often clearly aligned (Figures 7a and 7b). Large, multibead clumps may develop if the culture is left static for any length of time or is stirred at too slow a speed (Figure 7c). Maintaining the stirring speed at 40 to 50 rpm will minimize bead clumping. If the vessel wall is improperly treated with silicon or if any area of silicon has worn thin, the bridged microcarriers will become attached to the wall and can form long strands of cells, beads, and matrix. Two- and three bead clumps are found in almost all cultures regardless of their age but do not present a significant problem.

The "bridging" noted with SMC is in contrast to vascular endothelial cells, a highly contact inhibited cell with marked polarity of apical and basal surfaces. These cells, for example, are easily grown in completely static conditions, with little evidence of bead clumping (Figure 7d).

Older cultures which have laid down a large amount of extracellular matrix will sometimes shear off in sheets if the stirring speed is too great. The effect can be identified by the appearance of matrix or cells which can be seen dangling from partially or completely denuded beads. Pipetting the microcarriers through too small an opening will also produce shearing of cells. To avoid this, we routinely use wide-bore pipettes to dispense the colonized microcarriers. Any microcarrier culture will contain a few unpopulated beads even at their most mature state. These represent a tiny fraction of the commercial microcarriers that remain unattractive to cells, probably because of incomplete treatment of the bead surface during manufacture.

D. Scanning Electron Microscopy

Microcarriers can readily be processed for scanning electron microscopy (SEM). The use of solid beads minimizes shrinkage artifact, whereas porous beads are less suitable. After fixation in 2% glutaraldehyde in buffered salt solution, pH 7.4, for 1 hr at 4°C, the cells are washed twice with deionized water. All manipulations can be performed in a plastic conical centrifuge tube, with gentle inversion of the tube between each step. After the beads settle, the supernatant is removed and replaced with the next solution. Cell water is gradually exchanged with alcohol by replacement with ethanol solutions of increasing concentrations (10 to 100% at 10% increments for 10 min each step at room temperature). Microcarriers are then either critical-point dried from ethanol/CO_2 or air-dried from 100% ethanol, transferred to sticky copper tape mounted on an SEM stub, and coated with a thin layer (approximately 200 Å) of gold-palladium (Hummer IV sputter coater, Technics Instruments, Alexandria, Va.). An example of endothelial cells prepared by this procedure and viewed in an AMR 1000 scanning electron microscope is shown in Figure 8a.

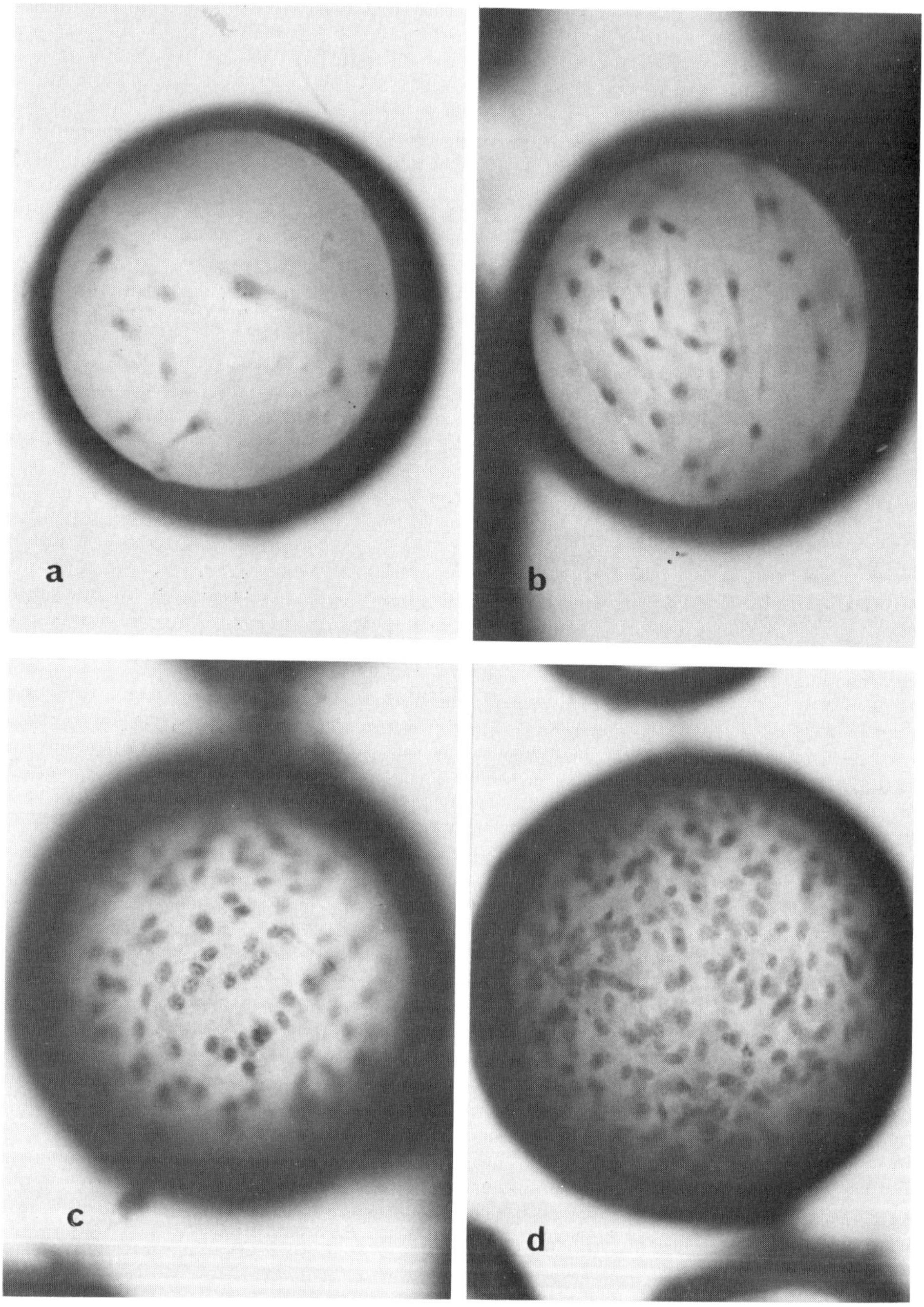

FIGURE 6. Vascular SMC stained with hemotoxylin at various cell densities on solid polystyrene microcarriers. Average cell-to-bead ratios in the cultures from which these examples were taken were (a) 30, (b) 60, (c) 110, (d) 155. (Magnification × 280.)

E. Transmission Electron Microscopy

The following procedure is a modification of that published by Sargent et al.[30] It was developed by Netland and Damm of the Department of Physiology and Biophysics, Harvard Medical School, and was kindly provided for this chapter by them.

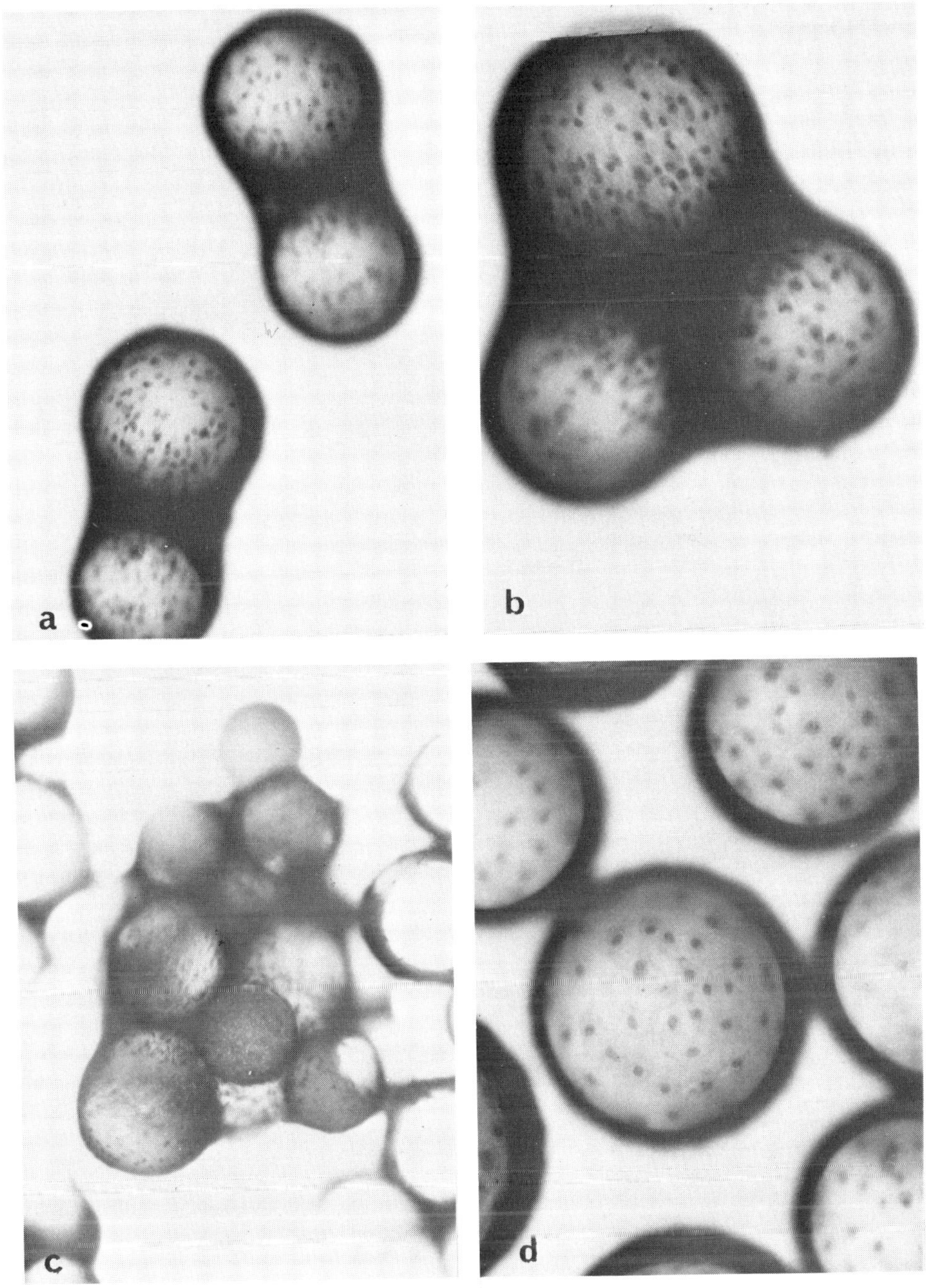

FIGURE 7. Examples of microcarrier aggregates which often occur in cultures of vascular SMC and fibroblasts. (a) Doublets, common in most SMC cultures even under optimal stirring conditions; magnification × 120. (b) Triplets, occasionally observed in normal SMC cultures; magnification × 155. (c) Extensive aggregation and clumping of SMC microcarriers in an insufficiently stirred culture. Cells in the interior of such aggregates will become necrotic; magnification × 90. (d) In contrast to SMC and fibroblasts, microcarrier cultures of vascular endothelial cells do not readily form aggregates; magnification × 185. Biosilon microcarriers. Hematoxylin stain.

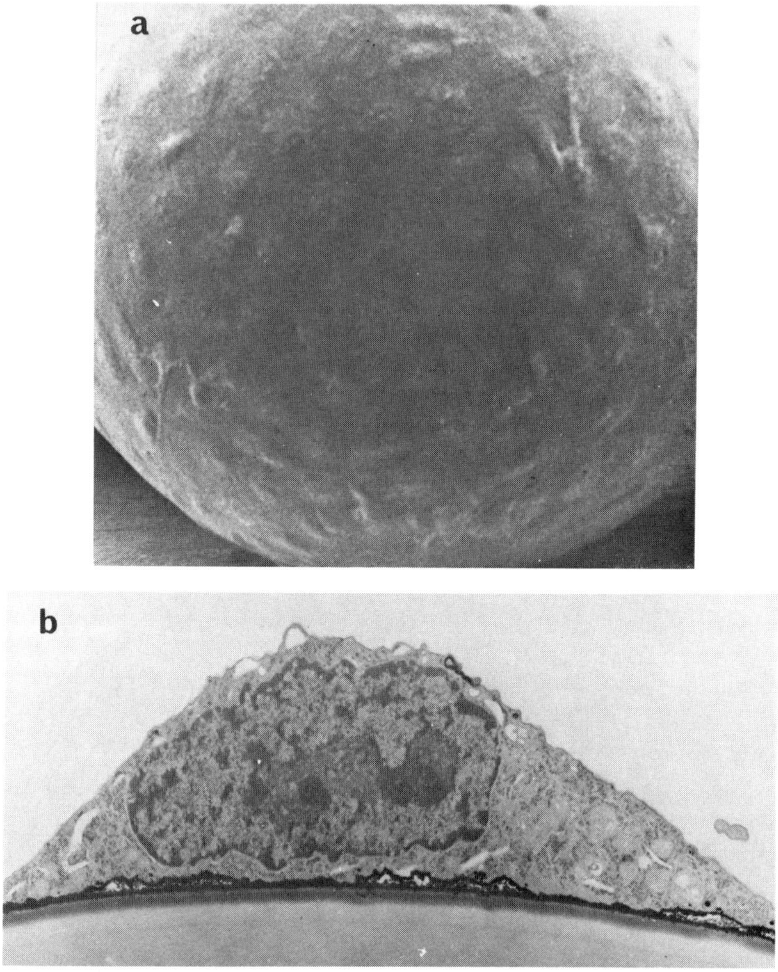

FIGURE 8. Electron microscopy of microcarrier culture (Biosilon) demonstrated for vascular endothelial cells. (a) Scanning electron microscopy of confluent bovine aortic endothelial cells; magnification × 500. (b) Transmission electron microscopy of same culture as in (a). Tannic acid was used postfixation to emphasize the interface between cell and bead surface; magnification × 20,000.

Following fixation and washing as described above for SEM, the microcarriers in buffered salt solution are warmed to 45°C in a water bath; 2% agar in PBS is prepared and also warmed to 45°C. A few drops of 1% eosin are added to make the agar visible during processing; 1 to 2 mℓ of the agar solution is added to the sample. The sample and agar solution are then gently mixed, using a wooden stick. The sample is centrifuged at 1000 rpm for 2 min. After the agar block is formed, the sample is removed by cutting the centrifuge tube and flooding with PBS. The excess agar is cut off the agar block. The block of agar containing the sample is diced into 1 mm cubes which are then transferred to a glass vial. Following postfixation for 1 hr in 0.2 *M* cacodylate buffer containing 1% osmium tetroxide, the cubes are washed in three changes (10 min each) of 0.2 *M* cacodylate buffer. The samples are then washed three times (5 min each) with distilled water and gently shaken in 0.2 *M* cacodylate buffer containing 1% (w/v) uranyl acetate. After 1 hr of staining, cubes are washed in three changes of distilled water and processed through graded ethanol solutions (50%, 10 min; 70%, 10 min; 95%, 10 min; 100%, 10 min; 100%, 15 min). The samples are then processed through three changes of 1,2 epoxypropane and finally embedded in resin.[31] The changeover from dehydration fluid to embedding fluid is made in three stages: 1/1 dehydrating/embedding fluid for a half hour; 1/3 dehydrating/embedding fluid for a half hour; 100% embedding fluid for 1 hr; 100% embedding fluid overnight. Samples are cured for at least 10 hr in a 70°C oven. Semithin sections are cut and stained with 0.1% toluidine blue in 1% sodium borate and examined with the light microscope. Ultrathin sections are prepared from selected areas, picked up on copper grids, and stained with 7.5% uranylacetate in 50% ethanol followed by lead citrate stain.[32] The sections can then be examined in the transmission electron microscope, usually at 60 KV. An example of Biosilon-bound endothelial cells prepared for transmission electron microscopy (TEM) is shown in Figure 8b.

F. Applications in Vascular Research

The most useful properties of microcarriers as substrata for anchorage dependent cells are the high surface area to volume ratio and the ease with which the cells can be transferred between different locations without enzymatic removal from the substratum. The latter property, in particular, has proved useful for studies of the biochemistry of cell surface receptors and other surface proteins whose integrity is destroyed by proteolytic enzymes such as trypsin. The introduction of naked beads to a stirring microcarrier culture of SMC results in their colonization and hence an effective increase in the cell population. Such a phenomenon may be useful for the continuous culture of cells, avoiding the use of trypsin.[33] The property of cell migration onto and off beads has been used to selectively isolate microvessel endothelial cells[34] and to "plate" cells into petri dishes without using trypsin, the cells simply migrating onto the dish surface. Busch et al.[15] and this laboratory[16,33] have set up perfusion beds of microcarrier-borne cells to collect, in concentrated form, metabolites secreted by the cells, thus using the high surface area to volume ratio of microcarriers to create a very small interstitial volume of medium.

Each bead is essentially a microculture of up to 200 anchorage-dependent vascular cells. This fact provides great versatility in moving different numbers of cells into and out of selected environments. We have used such an approach to study vascular cell interactions in co culture.[17,18,35,36] There are subtle but important differences between studies of intercellular metabolism in co-culture compared with the use of conditioned medium. Co-culture allows the potential for multiple interactions between the cell populations whereas there is an inherent restriction of interactions when conditioned medium is used. First, there is no provision for signals or messages from the target cell to reach the cell that is conditioning the medium. Second, there is the possibility that short-lived potent factors produced by the conditioning cell may be inactive either because of short half-life or by dilution when applied as conditioned medium to the target cell. Third, the separation of cell populations for the production of conditioned medium removes the normal checks and balances which presumably exist between cell types such as endothelium and SMC residing so close together in vivo. Co-culture minimizes these restrictions. Two examples of the usefulness of this approach have recently been reported.[18,35,36,37] We show in Figures 9a and 9b the two major co-culture systems used in this laboratory for vascular cells, one involving humoral interactions only and the other which allows both contact between the cell populations and humoral interactions to occur simultaneously.

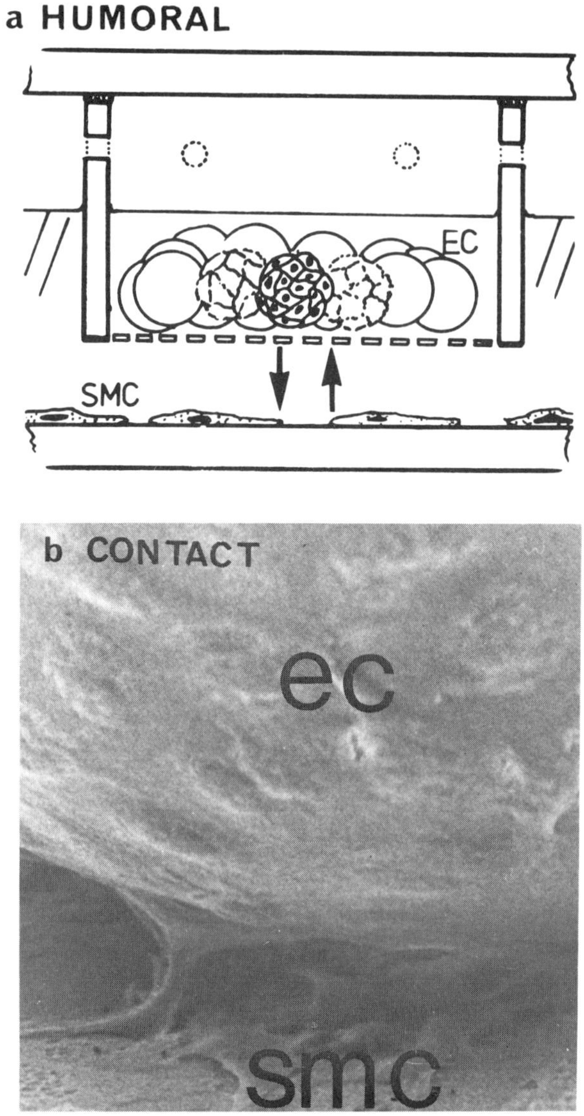

FIGURE 9. Two systems for the study of vascular cell interactions utilizing microcarrier cultures. (a) Humoral interactions between confluent endothelial cells on microcarriers and SMC grown in conventional petri dish culture. See References 17 and 18. (b) Contact co-culture between endothelial (ec) and smooth muscle cells (smc). In such a configuration, intercellular communication by both humoral diffusion and junctional transfer has been demonstrated. See References 35, 36, and 37.

VIII. CONCLUDING REMARKS

Microcarriers have now become a useful and highly versatile tool for small-scale (nonindustrial scale) cell research. Many innovative applications, in addition to those cited above, are ripe for investigation, including the possible use of microcarriers in the seeding of vascular grafts, the isolation of endothelial cells in poorly accessible regions of the microcirculation, and the assay of cellular metabolism in living cells in a cuvette by fluorescence spectrophotometry. Such approaches should provide insight into the role of SMC, endothelial cells, and monocyte-derived macrophages in vascular biology and pathology.

ACKNOWLEDGMENTS

This work is funded by grants to Dr. Davies from the National Heart Lung and Blood Institutes at the National Institutes of Health (HL24612 and HL22602). We gratefully acknowledge the following members of the group and collaborators involved in microcarrier technology and related biochemical studies during the past 5 years: Catherine Kerr, Lesley Kuczera, Barbara Eisenhaure, Debbie Sandstrom, Pip Diehl and Drs. Henry Warren, George Truskey, and Peter Ganz.

REFERENCES

1. van Wezel, A. L., Growth of cell strains and primary cells on microcarriers in homogeneous culture, *Nature (London)*, 216, 64, 1967.
2. Montagnon, B. J., Vincent-Falquet, J. C., and Fanget, B., Thousand liters scale microcarrier culture of vero cells for killed polio virus vaccine; promising results, *Dev. Biol. Stand.*, 55, 37, 1984.
3. van Wezel, A. L., van der Velden de Groot, C. A. M., and van Herwaarden, J. A. M., The production of inactivated polio vaccine on serially cultivated kidney cells from captive-bred monkeys, *Dev. Biol. Stand.*, 46, 151, 1980.
4. Meignier, B., Mougeot, H., and Favre, H., Foot and mouth disease virus production on microcarrier-grown cells, *Dev. Biol. Stand.*, 46, 249, 1980.
5. van Hemert, P., Kilburn, D. G., and van Wezel, A. L., Homogenous cultivation of animal cells for the production of virus and virus products, *Biotechnol. Bioeng.*, 11, 875, 1969.
6. Edy, V. G., Augenstein, D. C., Edwards, C. R., Cruttenden, V. F., and Lubiniecki, A. S., Large-scale tissue culture for human IFN-production, *Tex. Rep. Biol. Med.*, 41, 169, 1982.
7. Page, M. and Dufour, C., Growth of human colon carcinoma cells on beaded microcarriers, *J. Cell Biol.*, 83, 302a, 1979.
8. Griffiths, B., Atkinson, T., Electricwala, A., Latter, A., McEntree, I., Riley, P. A., and Sutton, P. M., Production of a fibrinolytic enzyme from cultures of guinea pig keratocytes grown on microcarriers, *Dev. Biol. Stand.*, 55, 31, 1984.
9. Prestidge, R. L., Sandlin, G. M., Koopman, W. J., and Bennett, J. C., Interleukin 1: production by P388D cells attached to microcarrier beads, *J. Immunol. Methods*, 46, 197, 1981.
10. Scott, R. W., Eaton, D. L., and Baker, J. B., Rapid purification of protease nexin and plasminogen activator released by human fibroblasts in microcarrier culture, *J. Cell Biol.*, 95, 121a, 1982.
11. Norrgren, G., Ebendal, T., Gelb, C., and Wikstrom, H., The use of Cytodex 3 microcarriers and reduced serum media for the production of nerve growth promoters from chicken heart cells, *Dev. Biol. Stand.*, 55, 43, 1984.
12. Thilly, W. G. and Levine, D. W., Microcarrier culture: a homogeneous environment for studies of cellular biochemistry, *Methods Enzymol.*, 58, 184, 1979.
13. Gotlib, L. J., Isolation of cell plasma membranes on microcarrier culture beads, *Biochim. Biophys. Acta*, 602, 207, 1980.
14. Ryan, U. S., Mortara, M., and Whitaker, C., Methods for microcarrier culture of bovine pulmonary artery endothelial cells avoiding the use of enzymes, *Tissue Cell*, 12, 619, 1980.
15. Busch, C., Cancilla, P. A., DeBault, L. E., Goldsmith, J. C., and Owen, W. G., Use of endothelium cultured on microcarriers as a model for the microcirculation, *Lab. Invest.*, 47, 498, 1982.
16. Davies, P. F., Microcarrier culture of vascular endothelium, *Exp. Cell Res.*, 134, 365, 1981.
17. Davies, P. F. and Kerr, C., Co-cultivation of vascular endothelial and smooth muscle cells using microcarrier techniques, *Exp. Cell Res.*, 141, 455, 1982.
18. Davies, P. F., Truskey, G. A., Warren, H. B., O'Connor, S. E., and Eisenhaure, B. H., Metabolic cooperation between vascular endothelial cells and smooth muscle cells in coculture, *J. Cell Biol.*, 101, 871, 1985.

19. Gebb, C., Clark, J. H., Hirtenstein, M. D., Lindgren, G., Lindskog, V., Lindgren, B., and Uretblad, P., Alternative surface for microcarrier culture of animal cells, *Dev. Biol. Stand.,* 50, 93, 1982.
20. Reuveny, S., Microcarriers for culturing mammalian cells and their applications, *Adv. Biotechnol. Processes,* 2, 1, 1983.
21. Pharmacia Fine Chemicals: *Microcarrier Cell Culture: Principle and Methods,* Trade Publication, 1982.
22. Johansson, A. and Nielsen, V., Biosilon, a new microcarrier, *Dev. Biol. Stand.,* 46, 125, 1980.
23. Hirtenstein, M. D., Clark, J. M., and Gebb, C., A comparison of various laboratory scale culture configurations for microcarrier culture of animal cells, *Dev. Biol. Stand.,* 50, 73, 1982.
24. DeBruyne, N. A. and Morgan, B. J., Stirrer for suspension cell cultures, *Am. Lab.,* June 1981.
25. Ross, R., The smooth muscle cell. II. Growth of smooth muscle in culture and formation of elastin fibers, *J. Cell Biol.,* 50, 172, 1971.
26. Chamley-Campbell, J. H., Campbell, G. R., and Ross, R., Phenotype-dependent response of cultured aortic smooth muscle to serum mitogens, *J. Cell Biol.,* 89, 378, 1981.
27. Cocks, T. M., Angus, J. A., Campbell, J. H., and Campbell, G. R., Release and properties of endothelium-derived relaxing factor (EDRF) from endothelial cells in culture, *J. Cell. Physiol.,* 123, 310, 1985.
28. Vogel, A. Raines, E., Kariya, B., Rivest, M.-J., and Ross, R., Coordinate control of 3T3 cell proliferation by platelet-derived growth factor and plasma components, *Proc. Natl. Acad. Sci. U.S.A.,* 75, 2810, 1978.
29. Lowry, D. H., Rosebrough, N. J., Farr, A. L., and Randall, R. J., Protein measurement with the folin phenol reagent, *J. Biol. Chem.,* 193, 265, 1951.
30. Sargent, G. F., Sims, T. A., and McNeish, A. S., The use of polystyrene microcarriers to prepare cell monolayers for transmission electron microscopy, *J. Microsc. (Oxford),* 122, 204, 1980.
31. Spurr, A. R., A low-viscosity epoxy embedding medium for electron microscopy, *J. Ultrastruct. Res.,* 26, 31, 1969.
32. Sato, J., Lead citrate stain in electron microscopy, *Electron Microsc.,* 17, 158, 1968.
33. Davies, P. F., Microcarrier cultures in vascular endothelial research, *Dev. Biol. Stand.,* 50, 125, 1982.
34. Folkman, J., Capillary endothelial cells — the use of microcarrier beads for isolation and long-term culture, *J. Cell Biol.,* 95, 485 A, 1982.
35. Davies, P. F., Ganz, P., and Diehl, T. S., Reversible microcarrier-mediated junctional communication between endothelial and smooth muscle cell monolayer: an in vitro model of vascular cell interactions, *Lab. Invest.,* 85, 710, 1985.
36. Ganz, P., Davies, P. F., Leopold, J., Gimbrone, M. A., and Alexander, R. W., Short and longterm interactions of endothelium and vascular smooth muscle in coculture: effects on cyclic GMP production, *Proc. Natl. Acad. Sci. U.S.A.,* 83, 3552, 1986.
37. Davies, P. F., Biology of disease — vascular cell interactions with special reference to the pathogenesis of atherosclerosis, *Lab. Invest.,* 55, 5, 1986.

Chapter 4

PHENOTYPIC MODULATION OF SMOOTH MUSCLE CELLS IN PRIMARY CULTURE

Gordon R. Campbell and Julie H. Campbell

TABLE OF CONTENTS

I. INTRODUCTION

Various cell types can alter their character when the environment changes. These alterations in character are called modulations of the differentiated state and involve reversible interconversions between phenotypes. Modulations in cell phenotype may occur as a result of a number of factors, such as cell interactions, alterations of extracellular matrix, or in response to other signals such as hormones.

Smooth muscle cells (SMC) are capable of expressing a range of phenotypes. At one end of this spectrum of phenotypes are the SMC whose function is almost exclusively that of contraction ("contractile state"). This function is reflected structurally with 80 to 90% of the cell cytoplasmic volume occupied with contractile apparatus. Organelles such as rough endoplasmic reticulum (ER), Golgi, and free ribosomes are few in number and located in the perinuclear region (Figure 1). Cells of the taenia coli, vas deferens, circular muscle of the ileum, and smaller muscular arteries express this phenotype. These muscles contain relatively little connective tissue (e.g., the volume of extracellular space is about 30% in taenia coli and 10 to 15% in the circular muscle of the ileum),[1] and in visceral muscle much of this connective tissue is synthesized by constituent fibroblasts. Therefore, SMC need to express minimal synthetic function. At the opposite end of the spectrum of phenotypic expression is the muscle cell, whose function is almost exclusively that of synthesis ("synthetic state") (Figure 2). As with other cell types actively engaged in the production of extracellular matrix (such as fibroblasts), the cytoplasm contains few filament bundles but large amounts of rough ER, Golgi, and free ribosomes. Cells in the synthetic state are seen in development, repair, and pathological conditions.[2]

It must be emphasized that the contractile and synthetic states as described above represent the extreme ends of a continuous spectrum of phenotypes, with most smooth muscle expressing intermediate morphology. For example, in large mammalian elastic arteries, smooth muscle is the only cell type present,[3-5] and 60 to 70% of the cell volume is occupied with myofilaments.[6,7] The volume of the connective tissue is 50 to 60%; therefore, this muscle cell has a dual function, being responsible for contraction/relaxation and the normal replacement of extracellular matrix components.

II. STRUCTURAL AND FUNCTIONAL CHANGES OF ARTERIAL MEDIAL SMOOTH MUSCLE IN PRIMARY CULTURE

A. Explant Cultures

Kasai and Pollak in 1964[8] were the first to describe phenotypic changes of SMC in culture. They examined the cellular outgrowth from rabbit arterial wall explants and suggested that "fibroblasts" appeared within 3 to 5 days and "smooth muscle cells" within 8 to 10 days. Similar observations were made on explants of human aorta in culture.[9] However, even special light microscope stains (such as Pollak's trichrome), are inadequate to distinguish the spectrum of phenotypes that SMC express. The electron microscope is far more appropriate for this task, and Dauod, Fritz, Jarmolych, and colleagues[10,11] described the presence of four "cell types" in outgrowth from pig aortic media explants. These were termed "primitive cells", "fibroblast-like cells", "modified smooth muscle cells", and "mature smooth muscle cells", according to the number of filaments and the concomitant increase or decrease in the number of synthetic organelles present. They suggested that mature SMC "dedifferentiated" in the first few days in culture to modified SMC, then to fibroblast-like cells, then finally to primitive cells, where they rapidly proliferated. Peak proliferation (as determined by ^{3}H-thymidine incorporation into DNA) occurred between 9 to 14 days in culture, whereas production of extracellular matrix (collagen, elastin, and glycosaminoglycans)

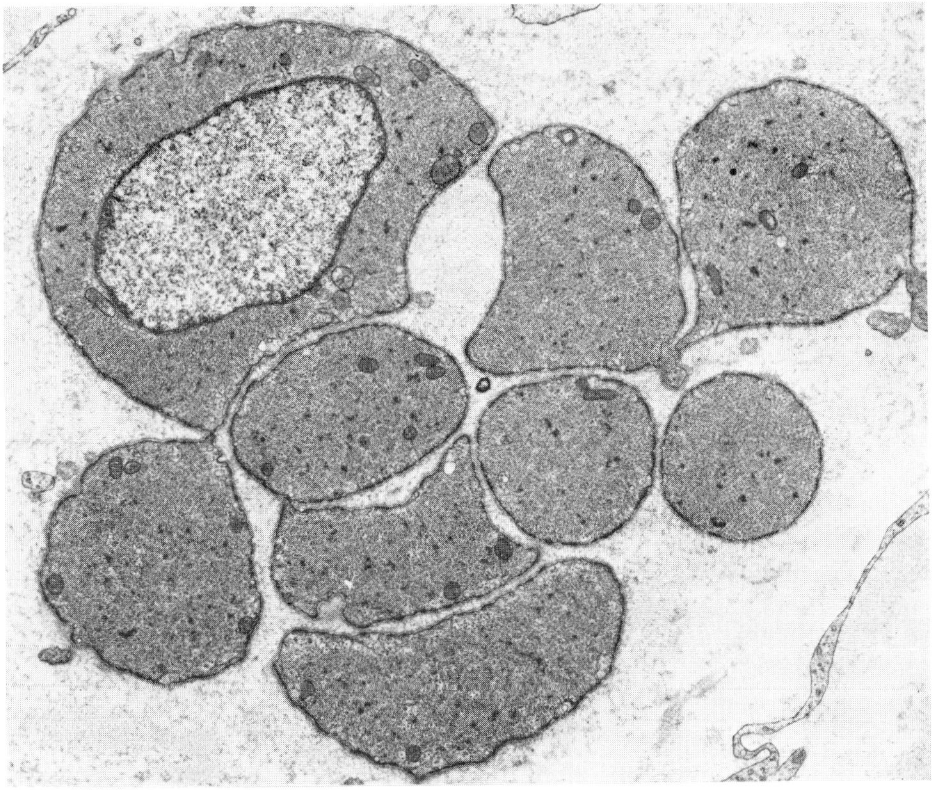

FIGURE 1. SMC from the mouse vas deferens. Most of the cytoplasm is filled with myofilaments. These cells could be considered to be in the "contractile state". (Magnification × 4500.)

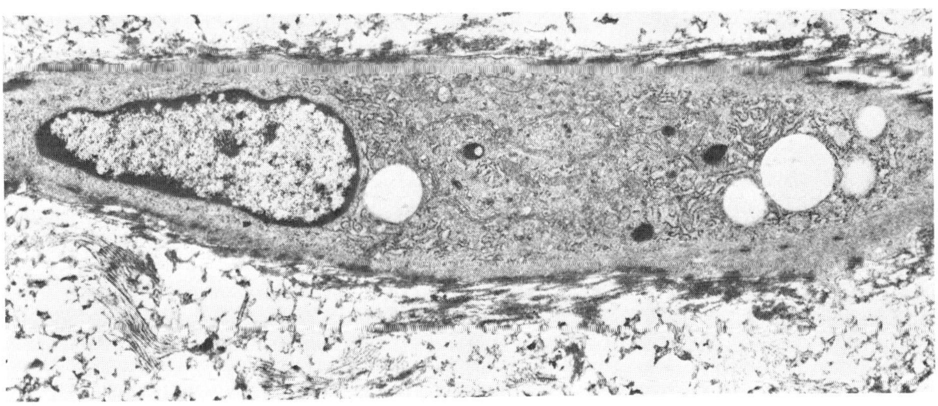

FIGURE 2. SMC from the intima of a hypercholesterolemic rabbit. Most of the cytoplasm is filled with organelles generally associated with synthesis. Myofilament bundles are small and located towards the periphery of the cell. This cell is towards the "synthetic state". (Magnification × 3000.)

was minimal at 7 to 9 days, but then increased up to 21 days. The primitive cells subsequently "redifferentiated" back to smooth muscle. These findings have more recently been substantiated by Rossi et al.[12] and Wight et al.[13]

B. Enzyme-Dispersed Cells

1. Terminology

When SMC from the guinea pig vas deferens and taenia coli are enzymatically dispersed and placed in primary culture, they undergo similar morphological changes to those described in Section A. These changes have been described as dedifferentiation and redifferentiation,[14-19] but it has since been recognized that this terminology is not strictly correct since there is no evidence that the more "primitive"-looking SMC return to a multipotential state (true dedifferentiation). The word "modulation" has subsequently been used to describe the changes which occur in smooth muscle phenotype in response to an altered environment.[2,20-23]

2. Changes in Morphology

Vascular and visceral smooth muscle enzyme-dispersed into single cells and plated in primary culture attach and flatten on the culture substratum within 1 to 2 days. They then closely resemble the same cells in vivo[20,21] (Figure 3). That is, their cytoplasm contains substantial bundles of myofilaments. They are therefore described as being in the contractile phenotype. If they are seeded below a critical cell concentration, the cells undergo a series of spontaneous changes in phenotype with time to what is termed the synthetic state.[16,19,21,22] In the synthetic state, SMC contain few filaments but increased numbers of organelles involved with synthesis (Figure 4). At the same time, there is a decrease or loss in staining reaction with FITC-labeled antibodies to the heavy meromyosin end of smooth muscle myosin.[19] The number of lysosomes also increases.[21] With phase-contrast microscopy, the cells change from being phase-dense and ribbon or fusiform in shape, to phase-lucent, broad, and flat (Figures 5 and 6). The time of phenotypic change is dependent on the donor species, age, and organ.[20,24] While phenotypic modulation from the contractile state to the synthetic state occurs in plasma-derived serum as well as in whole blood serum,[21,25] platelet factors can affect this process since platelet-derived-growth-factor (PDGF) antibodies delay the onset of phenotypic change in 10% whole blood serum, and purified PDGF speeds up the process in 10% plasma-derived serum.[21] Prostaglandin E1 also stimulates the transition of SMC from the contractile to the synthetic state during their first few days in culture. However, at a later time, when the majority of cells are in the synthetic state and have undergone one or more cell divisions, prostaglandin E1 slows down the rate of initiation of DNA synthesis.[26]

Thyberg et al.[21] have examined the cytoplasmic volume density of organelles in rat aortic smooth muscle with time in culture after enzyme dispersal and found that the volume percentage of filaments decreases from 60.32 on day 0 to 5.34 on day 4, while the volume percentage of rough endoplasmic reticulum increases from 5.33 on day 0 to 23.77 on day 4. Golgi cisternae also increases from 1.79 to 5.5 over the same time period, while lysosomal volume percent increases from 3.8 at day 0 to 6.53 on day 10.

3. Factors Affecting the Reversibility of Smooth Muscle Phenotypic Changes In Vitro

Smooth muscle phenotypic change is reversible or irreversible, depending on the initial cell seeding density.[24,27] If the cells are seeded at 10^6 cells per milliliter or greater, they form a confluent monolayer from the first day in culture. These cells do not undergo a spontaneous phenotypic change but remain in the contractile phenotype. If they are seeded sparser (i.e., at $5 \times 10^4 \rightarrow 1 \times 10^5$ cells per milliliter), they undergo the spontaneous phenotypic change on days 6 to 8, then proliferate until they form a confluent monolayer. If the cells undergo fewer than five cell doublings in forming this confluent monolayer, they will revert back to the contractile state within 1 to 2 days of confluency being reached. If, however, they have been seeded very sparsely (i.e., 1 ×

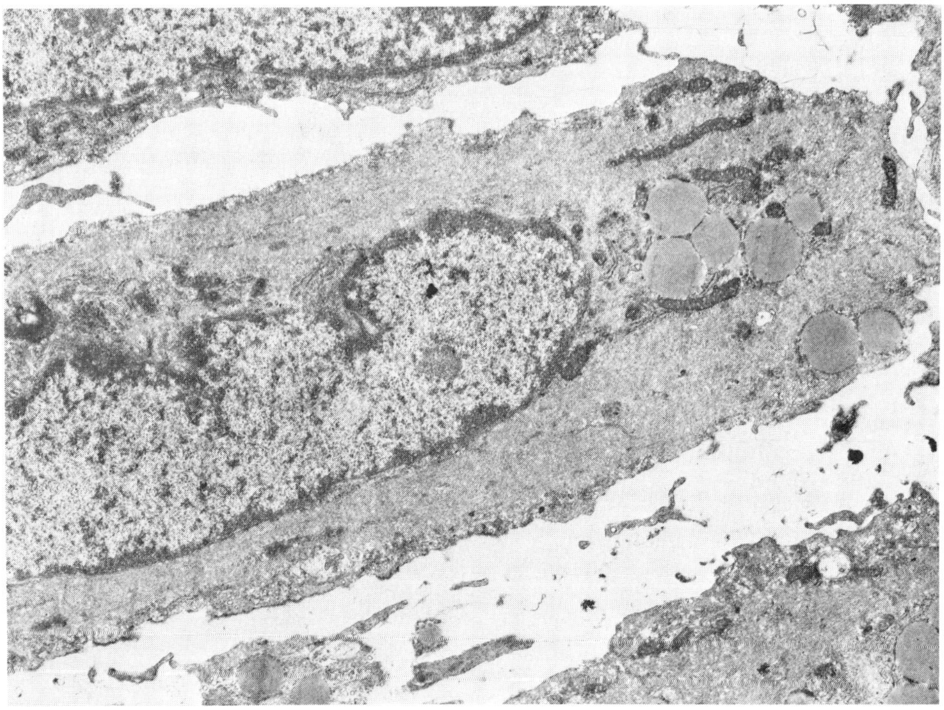

FIGURE 3. Seven-week-old rabbit aorta 4 days in primary culture. SMC have a similar appearance to those of the intact media, with much of their cytoplasm filled with myofilaments. (Magnification × 3000.) (From Schwartz, S. M., Campbell, G. R., and Campbell, J. H., *Circ. Res.*, 58, 427, 1986. With permission.)

$10^3 \rightarrow 5 \times 10^3$/mℓ), the cells will undergo the phenotypic change on days 6 to 8, but then need considerably greater than five cell doublings to achieve confluency. These cells do not return to the contractile phenotype, but remain permanently in the synthetic phenotype. SMC that are in the synthetic state but capable of returning to the contractile state have been termed "reversible synthetic". Cells which are incapable of returning to the contractile state have been termed "irreversible synthetic".[23] This loss of reversion of phenotype is not related to "senescence" demonstrated by Ross and Kariya[28] in late passage smooth muscle, since the primary cultured cells have undergone many fewer cell doublings and are still capable of intense proliferation. However, "irreversible synthetic" SMC after further cell doublings eventually become senescent and do not proliferate in response to serum mitogens.

4. Contraction

In the first few days in primary culture while the cells morphologically express a contractile phenotype, spontaneous contractions at a rate of one to eight per minute have been reported in single SMC from the guinea pig vas deferens,[14,16,17,29-31] taenia coli,[31-33] ureter,[18,31] and chicken gizzard.[15,31,34,36] There are a few reports of spontaneous contractions in single vascular SMC in culture[19,36] and of their stimulation to contract by vasoactive agents.[19,37]

Concomitant with the change in phenotype to a synthetic state, the muscle cells lose their ability for spontaneous or induced contraction.[20]

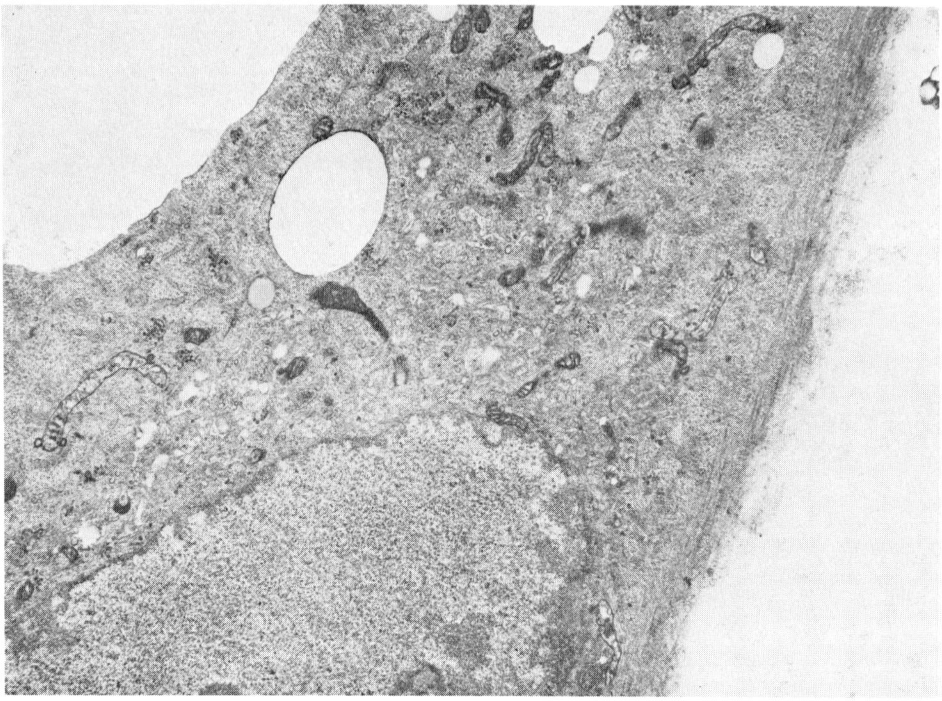

FIGURE 4. Seven-week-old rabbit aorta 11 days in culture (first subculture). Although a filament bundle extends along the right-hand side of the cell (note plasmalemmal vesicles of cell membrane), most of the cytoplasm is filled with organelles associated with synthesis. (Magnification × 3500.) (From Schwartz, S. M., Campbell, J. H., and Campbell, G. R., *Circ. Res.*, 58, 427, 1986. With permission.)

5. Division

Only 3 to 5% of aortic SMC from the adult pig and monkey in the first 6 or 7 days of primary culture incorporate ^{3}H-thymidine into their nuclei during a 4-hr, pulse-labeling period in the presence of 5% whole blood serum.[25,38] However, on day 7, 8, or 9 (i.e., the day after phenotypic modulation to the synthetic state is first evident morphologically), the number of labeled nuclei increases to 10 to 20% (see also Chapter 13). The percentage of cells incorporating ^{3}H-thymidine into DNA then increases linearly with time until on day 14, 40 to 60% of cell nuclei are labeled during the 4-hr, pulse-labeling period. The cells begin to proliferate logarithmically in 5% whole blood serum about 24 hr after they show an increased incorporation of ^{3}H-thymidine. There is no significant increase in cell number while the cells are in the contractile state; however, some contractile SMC are capable of division.[29]

The time of phenotypic change and subsequent proliferation is dependent upon the donor species, age, and organ. Thyberg et al.[21] studied isolated SMC from 8-month-old rat aorta arteries and found that by day 3 structural modification had occurred and cells began to proliferate logarithmically on day 5.

Phenotypic modulation occurs in the absence of the PDGF[21,25] and in the presence of mitomycin C[21] but in these cases is not followed by cell division. This demonstrates that while modulation is an essential but not sufficient prerequisite for SMC proliferation, the PDGF and/or other serum factors are necessary for mitosis to occur.

Recent observations by Nilsson et al.[39] show that, at least in the rat, modulation of SMC from the contractile to the synthetic phenotype is accompanied by the production of a PDGF-like protein and autocrine initiation of DNA synthesis. Serum-free condi-

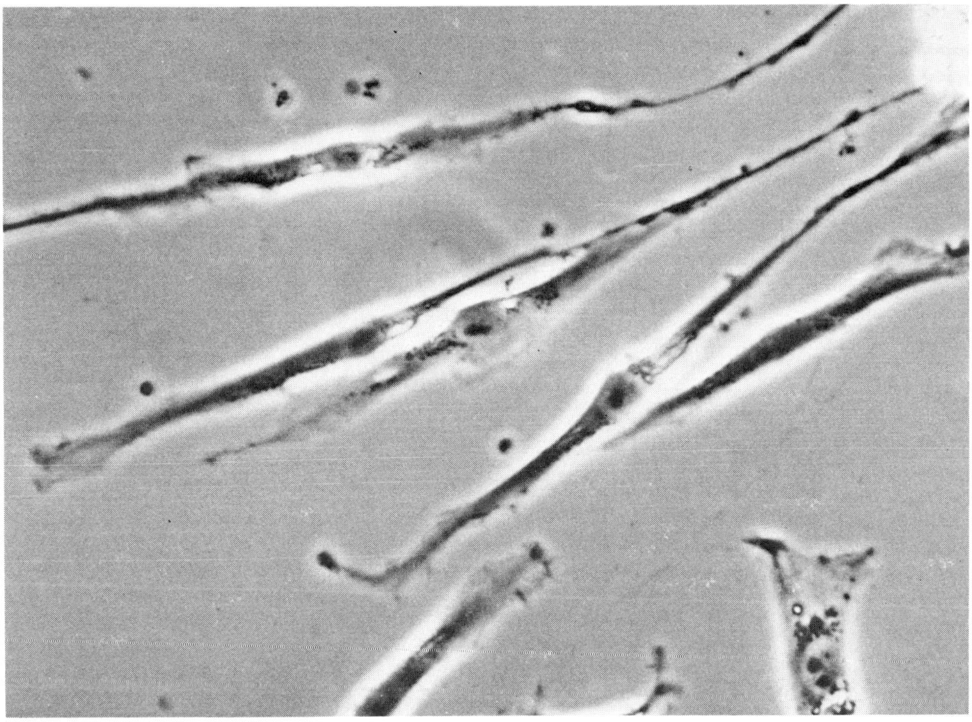

FIGURE 5. Enzyme-dispersed SMC from the adult pig aortic media, 3 days in primary culture. (Magnification × 800.)

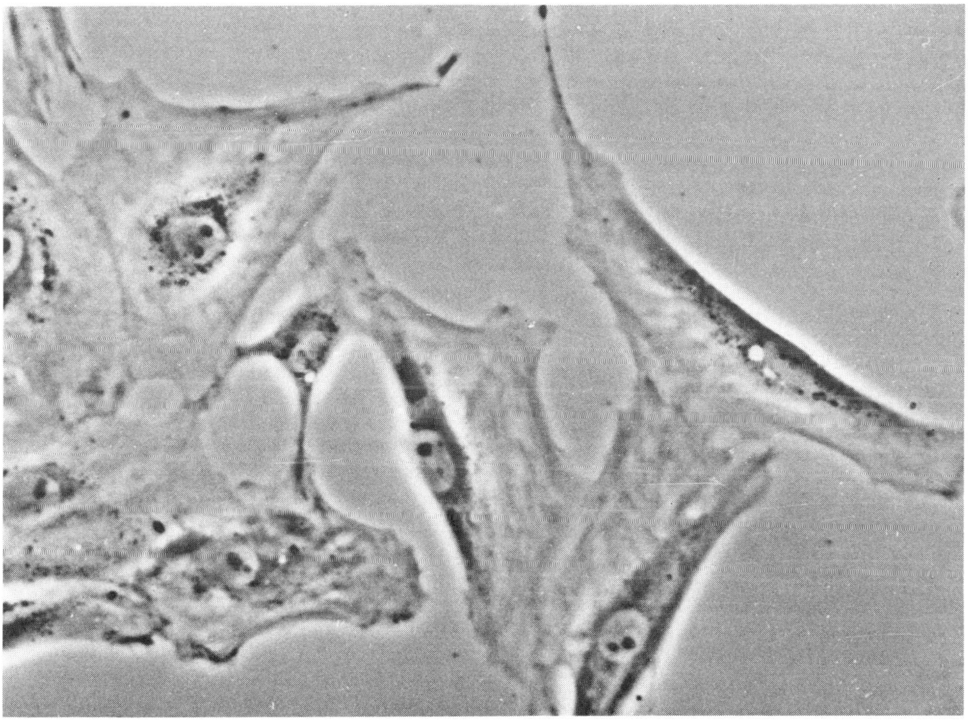

FIGURE 6. Enzyme-dispersed SMC from the adult pig aortic media, 9 days in primary culture. (Magnification × 800.)

tioned medium from primary cultures of adult rat arterial SMC contains a factor that initiates DNA synthesis in growth-arrested subcultures of SMC. The mitogenic activity can be neutralized by antibodies to PDGF, and no mitogenic activity occurs in conditioned medium from cultures pretreated with actinomycin D, excluding release into the medium of PDGF adsorbed to the plastic vessels during the initial culture in serum-containing medium. Exposure of human fibroblasts to samples of the conditioned media at 4°C inhibits subsequent binding of ^{125}I-labeled PDGF. Furthermore, SMC in primary cultures are able to initiate DNA synthesis in a chemically defined medium lacking PDGF and other growth factors.[39]

6. Synthesis of Extracellular Matrix

Synthetic state aortic SMC under basal conditions synthesize fourfold the amount of collagen as contractile state cells, as measured by the incorporation of ^{3}H-proline into hydroxyproline (see Reference 23). The amount of noncollagen protein synthesized doubles under the same conditions. Synthetic state cells also synthesize fivefold the contractile state level of sulfated glycosaminoglycans, as measured by uptake of inorganic sulfur-35.[23] These increases are not related to cell proliferation since in these experiments the synthetic state cells were maintained in a quiescent growth state.

7. Enzyme Alterations

Concomitant with the increased amount of synthetic organelles in the synthetic state SMC,[21] the specific activities of enzymes associated with mitochondria and rough endoplasmic reticulum increase.[40] Cytochrome C oxidase and NADPH-dependent cytochrome C reductase increase 2.3 and 1.8-fold, respectively. Increases of three- to fourfold in activity of the lysosomal enzymes acid phosphatase and *N*-acetyl-β-glucosaminidase occur,[40] consistent with the larger numbers of lysosomes present in synthetic state cells.[21] However, acid cholesteryl esterase, which is also found in lysosomes and is responsible for the hydrolysis of cholesteryl esters, decreases slightly but not significantly upon alteration to the synthetic state.[40]

8. Lipoprotein Metabolism

Cultured SMC in the synthetic state have an altered ability to metabolize lipoproteins.[38,40,41] Morphological examination of rabbit aortic SMC, incubated for 24 hr in 75 μg/mℓ β-very-low-density lipoprotein (β-VLDL) from the serum of cholesterol-fed rabbits, shows few lipid droplets, as demonstrated by Fettrot staining in "contractile"-state cells, with some increase in "reversible synthetic"-state cells, but with substantial numbers of lipid droplets in "irreversible synthetic"-state cells (Figures 7, 8, 9). This pattern of differential cholesteryl ester accumulation was confirmed by gas chromatography and by incorporation of [^{3}H] sodium oleate into cholesteryl oleate. "Irreversible synthetic"-state cells have a considerably greater capacity to bind and degrade β-VLDL than "reversible synthetic" cells, which in turn have a greater capacity than "contractile"-state cells (Figure 10). With all phenotypes the binding is receptor-mediated, of high affinity, and saturable. Similar results were found with normal VLDL, but the extent of the binding, degradation, and cholesteryl ester accumulation was lower than with β-VLDL. In contrast, cells of all three phenotypes bind and degrade considerably less low-density lipoprotein (LDL) under identical conditions, and the cells do not accumulate appreciable cholesteryl ester. Thus at least in the rabbit, lipoproteins of the triglyceride-rich spectrum (β-VLDL and VLDL) are more atherogenic than those of the cholesterol-rich spectrum (LDL), with "irreversible synthetic"-state cells more susceptible to lipid accumulation than "reversible synthetic"-state cells, which in turn are more susceptible than "contractile"-state cells.[41]

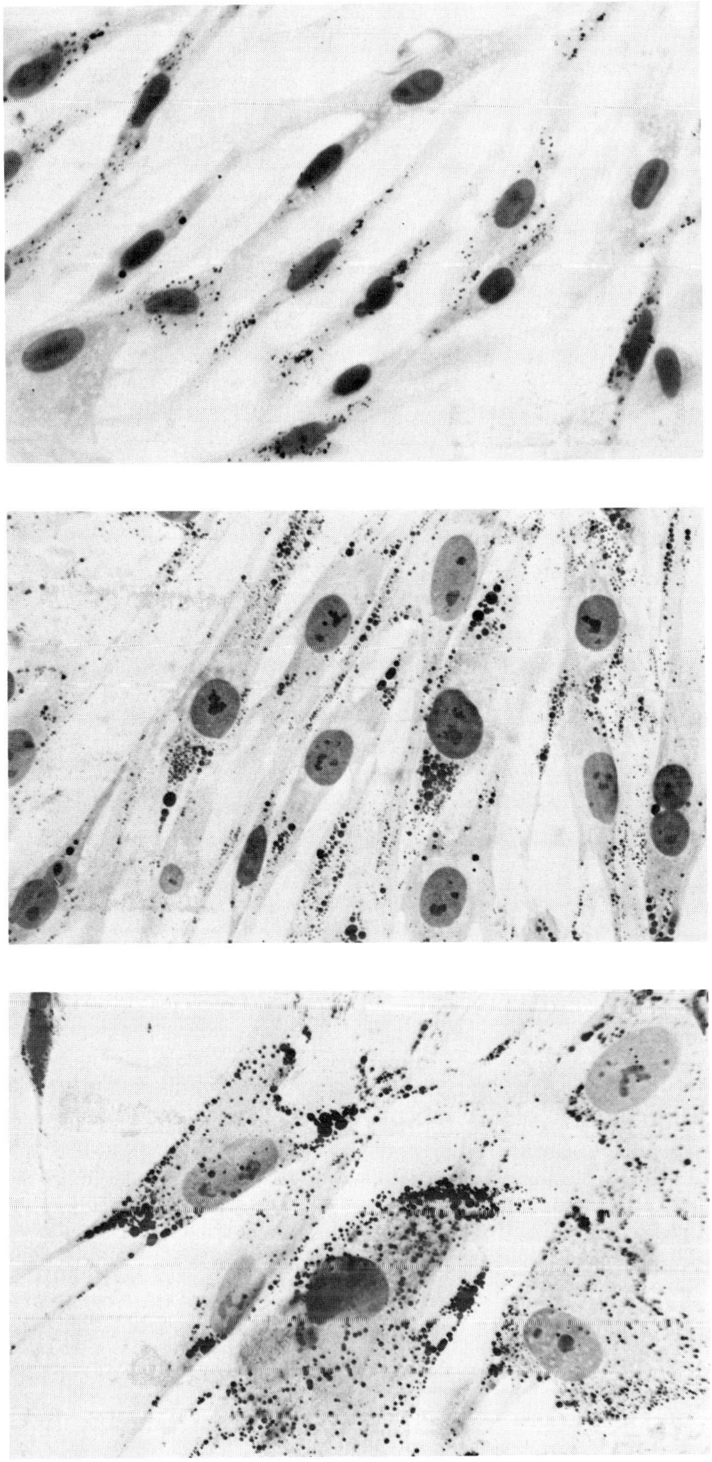

FIGURES 7, 8, and 9. Contractile (top, 7), reversible synthetic (middle, 8), and irreversible synthetic (bottom, 9) state cells stained with Fettrot solution for lipid droplets and counterstained with Meyer's Haemalum following 17-hr incubation in 5% LDS, then 24-hr incubation with 75 μg/mℓ β-VLDL in 5% LDS. (Magnification × 600.) (From Campbell, J. H., Reardon, M. F., Campbell, G. R., and Nestel, P. J., *Arteriosclerosis*, 5, 318, 1985. With permission.)

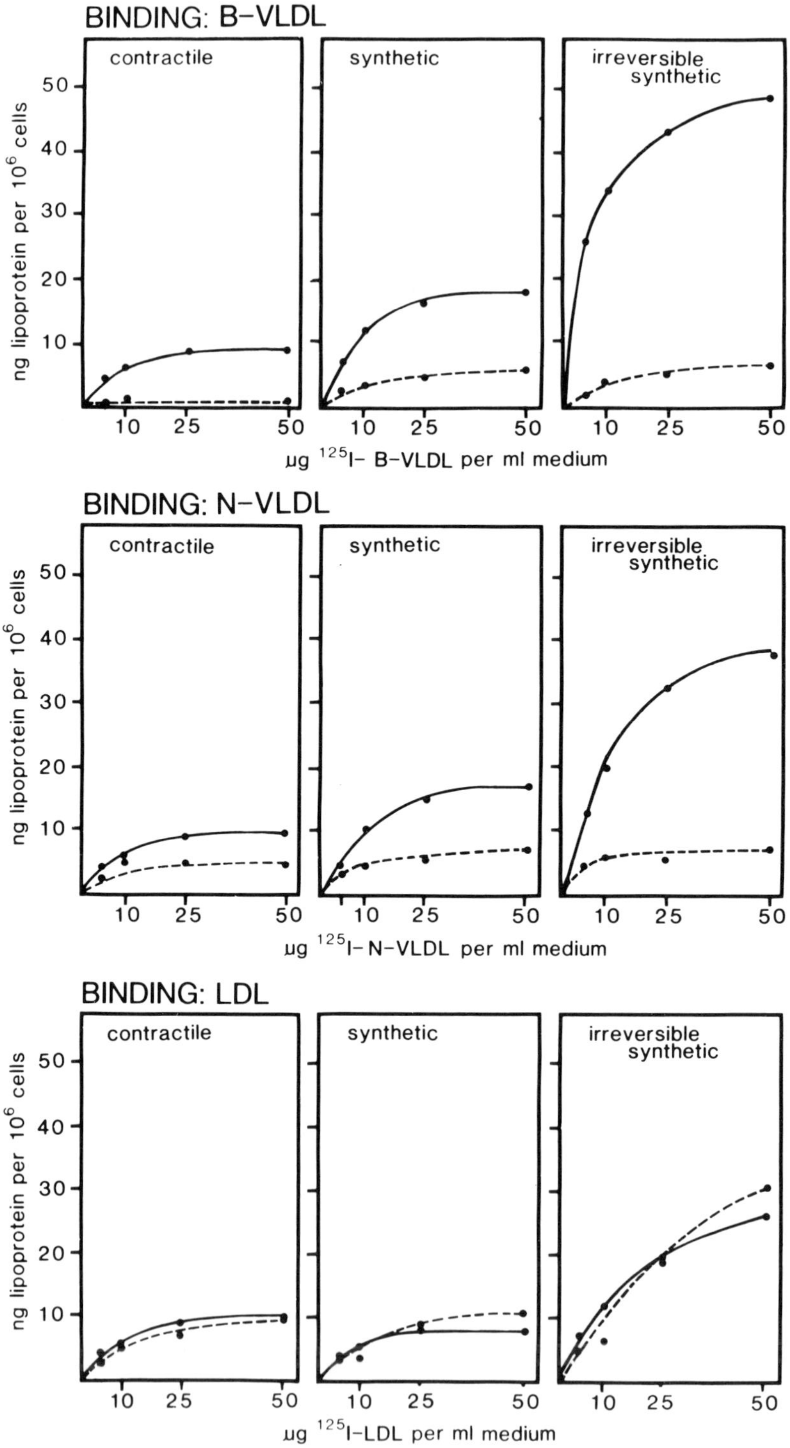

FIGURE 10. Specific (• — •) and nonspecific (O — O) binding of ^{125}I-labeled lipoproteins to rabbit aortic SMC in culture. (From Campbell, J. H., Reardon, M. F., Campbell, G. R., and Nestel, P. J., *Arteriosclerosis*, 5, 318, 1985. With permission.)

III. INFLUENCES OF ENDOTHELIUM ON SMOOTH MUSCLE PHENOTYPIC MODULATION AND PROLIFERATION

A. Influences on Smooth Muscle Phenotype

Spontaneous phenotypic modulation of enzymatically isolated smooth muscle in primary cell culture is prevented if the cells are grown in co-culture with a confluent monolayer of endothelial cells.[2,16,24,17] For this effect to occur, the two cell layers do not have to be in contact, but bathed by the same nutrient medium. The system used for these experiments comprises the modified Rose chamber, where the test and feeder cell layers are grown on opposite coverslips.[27] A feeder layer of confluent, contractile SMC (which is above the critical initial cell seeding density of 10^6/mℓ and does not undergo the spontaneous phenotypic change) also prevents the test, sparsely-seeded, contractile SMC from undergoing a spontaneous change in phenotype. In contrast, coculture with confluent adventitial fibroblasts, confluent and quiescent subcultured synthetic state SMC or subconfluent proliferating endothelial cells does not prevent the phenotypic modulation of the test contractile smooth muscle.[24]

The insertion of Spectrapor wet cellulose dialysis membranes of different molecular-weight cutoff between the feeder and test layers has demonstrated that the approximate molecular weight of the inhibitory substance or substances produced by confluent endothelial cells and contractile SMC is 10,000 to 15,000.[27] This is in agreement with our earlier studies which demonstrated that a glycosaminoglycan extract of about 12,000 daltons from the artery wall inhibits smooth muscle phenotypic change.[24] Treatment of this aortic glycosaminoglycan extract with four units per milliliter heparinase from *Flavobacterium heparinum* for 6 hr at 37°C destroys the active factor, suggesting that it is heparin or a related glycosaminoglycan. Addition of sodium heparin to culture medium also inhibits smooth muscle phenotypic change and subsequent proliferation.[24,27,42,43]

Cultured endothelial cells secrete about 50% of their glycosaminoglycan as a highly sulfated (heparin-like) heparan sulfate[44-47] and also produce a factor or factors able to stimulate glycosaminoglycan synthesis by SMC.[48] Heparan sulfate is the predominant glycosaminoglycan in the pericellular coat immediately adjacent to the plasma membrane of cultured SMC,[43,49,50] but by the seventh subculture, this is replaced by dermatan sulfate as the predominant species.[50]

B. Influences on Smooth Muscle Proliferation

Simms and Stillman in 1936[51] first discovered that the aorta contains growth inhibitory substances. Florentin et al.[52] showed that intraperitoneal injections of alcohol-precipitable components of an aqueous extract of pig aorta into 8 week-old living swine, in quantities as low as 500 μg of protein, drastically reduces the rate of entry of arterial SMC into mitosis, with no demonstrable effect on epidermal mitosis. Alcohol precipitates from aqueous extracts of pig aorta can be fractionated to produce both stimulators and inhibitors of synthetic state smooth muscle in culture.[53]

In culture, endothelial cells produce both stimulators and inhibitors of smooth muscle proliferation. The growth stimulator is called the endothelial-derived growth factor (EDGF),[54] and is mitogenic for many mesenchymally derived cells but not for endothelial cells themselves.[54-58] EDGF is a polypeptide which is not homogeneous and contains one factor that is similar to PDGF. This PDGF-like factor accounts for about 25% of the mitogenic effect, at least with 3T3 cells as the target.[59-61] The proliferative state of the endothelial cell determines whether a stimulator or inhibitor is released or which one predominates. When synthetic state smooth muscle is the test tissue and proliferating endothelium the feeder layer in a co-culture situation, the rate of proliferation of the test smooth muscle is stimulated. But when the feeder layer consists of confluent, quiescent endothelium, the normal growth of the test synthetic state smooth muscle is totally inhibited. A feeder layer of confluent contractile smooth muscle inhibits the proliferation of synthetic state cells, while a feeder layer of proliferating synthetic state cells sometimes stimulates, and other times inhibits, smooth muscle growth[24,27] (Table 1).

Similarly, conditioned medium containing 20% whole blood serum from confluent endothelial cells inhibits the proliferation of subcultured SMC, while conditioned medium from proliferating endothelial cells and from proliferating or confluent subcultured SMC and fibroblasts do not inhibit SMC growth.[55] The inhibitory activity has been amply demonstrated to be due to production of heparin-like glycosaminoglycans[43,55,62-64] (see also Chapter 7 for full discussion). Furthermore, heparan sulfate isolated from the surface of postconfluent SMC has recently been shown to exhibit a specific antiproliferative activity on cultured SMC 13 times greater than mucopolysaccharide obtained from the surface of exponentially growing SMC and more than 40 times greater than commercially available heparin.[65]

C. Possible Effects In Vivo

In the normal arterial wall, the SMC and abluminal surface of the endothelium are surrounded by a basal lamina consisting of (among other substances) laminin, heparan sulfate proteoglycan, and type IV collagen.[66,67] The basal lamina is, in turn, closely associated with the continuum of collagen, elastin, and proteoglycans of the intercellular matrix. Their association not only establishes the spatial arrangements of components within the vessel wall,[68] but may be responsible for maintaining SMC in the contractile phenotype, thereby inhibiting cell division (see Reference 69). Following endothelial denudation, platelets are deposited on the subendothelium, releasing PDGF from their alpha granules and platelet heparitinase (endoglycosidase) from their lysosomes. The endoglycosidase may cleave the heparin-like substance from the basal lamina proteoglycans of the SMC closest to the damaged lumen and from remaining endothelial cells and then be removed[62,70] (see Chapter 7). If a heparin-like glycosaminoglycan is the low-molecular-weight substance in the artery wall that maintains smooth muscle in the contractile phenotype, then the loss of this substance would result in SMC modulating to the synthetic phenotype (in primary cell culture it is due to dilution of the factor with decreasing cell seeding concentration). The synthetic-state muscle cells are responsive to substances from the blood and/or regenerating endothelium and/or macrophages, and the cells migrate into the intima and proliferate. Upon restitution of the endothelial lining, the confluent, quiescent endothelium and surrounding contractile-state SMC release the growth inhibitor for synthetic-state cells, which soon builds to a critical level, and development of the myointimal thickening is halted. The SMC then return to the contractile state (see also Chapter 7).

The notion that the extracellular matrix components can control the phenotype a cell expresses is not new. For instance, recent studies have shown that collagen type II and certain proteoglycans are influential in chondrogenic differentiation and in the regulation of chondrocyte phenotype.[71-73]

IV. SIGNIFICANCE OF SMOOTH MUSCLE PHENOTYPIC MODULATION IN PRIMARY CULTURE TO THE PATHOGENESIS OF ATHEROSCLEROSIS

Diffuse intimal thickenings of humans, like the fibromusculo-elastic intimal thickenings formed in experimental animals as a result of endothelial injury, are sites of predilection for atheromatous plaque formation but do not invariably develop into

Table 1
INFLUENCE ON SYNTHETIC STATE SMC PROLIFERATION OF CELL FEEDER LAYERS AFTER 5 DAYS IN CO-CULTURE

Feeder layer[a]	No. SMC/chamber × 10^5 ± S.D.
None (control)	20.6 ± 2.4
Confluent EC	0.9 ± 0.1 ($p < 0.001$)
Proliferating EC	25.2 ± 0.3 ($0.01 < p < 0.05$)
Confluent contractile SMC	1.7 ± 0.2 ($p < 0.001$)
Proliferating synthetic SMC	19.5 ± 6.5 ($p < 0.05$)

Note: Number of cells per chamber on day 0 is 0.8 ± 0.1 × 10^5.

[a] EC, endothelial cells; SMC, smooth muscle cells.

plaques.[74-77] Two mechanisms have been suggested by which diffuse intimal thickenings develop: (1) a response of the vessels wall to repeated injury and (2) a response of the vessel wall to postnatal growth and increase in blood pressure.[78] The nodular proliferation of SMC and accumulation of foam cells and extracellular lipid results in a reorganization of these thickenings in atherogenesis.

The vast majority of cells in human diffuse intimal thickenings are smooth muscle. These, like the SMC in lesions, vary in size and shape.[76,79-81] The volume fraction of synthetic organelles in SMC from human carotid endarterectomy and autopsy material has been measured by morphometric techniques,[82,83] and while intimal SMC from unaffected regions approximate those of the media, there is a 100% increase in volume fraction of synthetic organelles in SMC of diffuse intimal thickenings near lesions, as compared with those of the media. That is, the cells morphologically express the synthetic phenotype. Consistent with an altered phenotypic expression are the recent observations of Gabbiani et al.[84] that, unlike the aortic media which shows a predominance of actin in the α-isoform, human atheromatous plaques and surrounding adjacent diffuse intimal thickenings show a predominance of β-actin with some γ-isoform. Some of the SMC present in the diffuse intimal thickenings near lesions may be senescent, having exhausted their proliferative capacity during the formation of the thickening,[28,85] but others may actively express the same characteristics as the morphologically equivalent synthetic state SMC in culture. These cells respond rapidly to mitogens (whether they are from the serum, proliferating endothelium, or macrophages), synthesize substantial amounts of collagen, and exhibit enhanced binding and degradation of atherogenic lipoproteins, resulting in a substantial lipid accumulation,[38,40,41,69,76] all of which are features essential to the process of atherogenesis.

REFERENCES

1. Gabella, G., Structural apparatus for force transmission in smooth muscles, *Physiol. Rev.*, 64, 455, 1984.
2. Campbell, G. R., Chamley-Campbell, J. H., and Burnstock, G., Differentiation and phenotypic modulation of arterial smooth muscle cells, in *Structure and Function of the Circulation*, Vol. 3, Schwartz, C. J., Werthessen, N. T., and Wolf, S., Eds., Plenum Press, New York, 1981, 357.
3. Pease, D. C. and Paule, W. J., Electron microscopy of elastic arteries. The thoracic aorta of the rat, *J. Ultrastruct. Res.*, 3, 469, 1960.

4. Karrer, H. E., An electron microscope study of the aorta in young and in aging mice, *J. Ultrastruct. Res.*, 5, 1, 1961.
5. Paule, W. J., Electron microscopy of the newborn rat aorta, *J. Ultrastruct. Res.*, 8, 219, 1963.
6. Gerrity, R. G. and Cliff, W. J., The aortic tunica media of the developing rat. I. Quantitative stereologic and biochemical analysis, *Lab. Invest.*, 32, 585, 1975.
7. Olivetti, G., Anversa, P., Melissari, M., and Loud, A. V., Morphometry of medial hypertrophy in the rat thoracic aorta, *Lab. Invest.*, 42, 559, 1980.
8. Kasai, T. and Pollak, O. J., Smooth muscle cells in aortic cultures of untreated and cholesterol-fed rabbits, *Z. Zellforsch. Mikrosk. Anat.*, 62, 743, 1964.
9. Pollak, O. J. and Adachi, M., Human and rabbit arterial cells compared in tissue cultures, *Prog. Biochem. Pharmacol.*, 4, 294, 1968.
10. Jarmolych, J., Daoud, A. S., Landau, J., Fritz, K. E., and McElvene, E., Aortic media explants. Cell proliferation and production of mucopolysaccharides, collagen, and elastic tissue, *Exp. Mol. Pathol.*, 9, 171, 1968.
11. Fritz, K. E., Jarmolych, J., and Daoud, A. S., Association of DNA synthesis and apparent dedifferentiation of aortic smooth muscle cells *in vitro*, *Exp. Mol. Pathol.*, 12, 354, 1970.
12. Rossi, G. L., Alroy, J., and Röthenmund, S., Morphological studies of cultured swine aorta media explants, *Virchows Arch. B*, 12, 133, 1973.
13. Wight, T. N., Cooke, P. H., and Smith, S. C., An electron microscopic study of pigeon aorta cell cultures. Cytodifferentiation and intracellular lipid accumulation, *Exp. Mol. Pathol.*, 21, 1, 1977.
14. Chamley, J. H., Campbell, G. R., and Burnstock, G., An analysis of the interactions between sympathetic nerve fibres and smooth muscle cells in tissue culture, *Dev. Biol.*, 33, 344, 1973.
15. Campbell, G. R., Chamley, J. H., and Burnstock, G., Development of smooth muscle cells in tissue culture, *J. Anat.*, 117, 295, 1974.
16. Chamley, J. H., Campbell, G. R., and Burnstock, G., Dedifferentiation, redifferentiation and bundle formation of smooth muscle cells in tissue culture: the influence of cell number and nerve fibres, *J. Embryol. Exp. Morphol.*, 32, 297, 1974.
17. Chamley, J. H. and Campbell, G. R., Trophic influences of sympathetic nerves and cyclic AMP on differentiation and proliferation of isolated smooth muscle cells in culture, *Cell Tissue Res.*, 161, 497, 1975.
18. Chamley, J. H. and Campbell, G. R., Isolated ureteral smooth muscle cells in culture. Including their interaction with intrinsic and extrinsic nerves, *Cytobiologie Z. Exp. Zellforsch.*, 11, 358, 1975.
19. Chamley, J. H., Campbell, G. R., McConnell, J. D., and Groschel-Stewart, U., Comparison of vascular smooth muscle cells from adult human, monkey and rabbit in primary culture and in subculture, *Cell Tissue Res.*, 177, 503, 1977.
20. Chamley-Campbell, J., Campbell, G. R., and Ross, R., The smooth muscle cell in culture, *Physiol. Rev.*, 59, 1, 1979.
21. Thyberg, J., Palmberg, L., Nilsson, J., Ksiazek, T., and Sjolund, M., Phenotype modulation in primary cultures of arterial smooth muscle cells. On the role of platelet-derived growth factor, *Differentiation*, 25, 156, 1983.
22. Thyberg, J., Nilsson, J., Palmberg, L., and Sjolund, M., Adult human arterial smooth muscle cells in primary culture. Modulation from contractile to synthetic phenotype, *Cell Tissue Res.*, 239, 69, 1985.
23. Campbell, J. H. and Campbell, G. R., Chemical stimuli of the hypertrophic response to smooth muscle, in *Hypertrophic Response of Smooth Muscle*, Seidel, C. L., Ed., CRC Press, Boca Raton, Fla., 1987, in press.
24. Chamley-Campbell, J. H. and Campbell, G. R., What controls smooth muscle phenotype? *Atherosclerosis*, 40, 347, 1981.
25. Chamley-Campbell, J. H., Campbell, G. R., and Ross, R., Phenotype-dependent response of cultured aortic smooth muscle to serum mitogens, *J. Cell Biol.*, 89, 379, 1981.
26. Sjölund, M., Nilsson, J., Palmberg, L., and Thyberg, J., Phenotype modulation in primary cultures of arterial smooth-muscle cells. Dual effect of prostaglandin E1, *Differentiation*, 27, 158, 1984.
27. Campbell, J. H. and Campbell, G. R., Cellular interactions in the artery wall, in *The Peripheral Circulation*, Hunyor, S., Ludbrook, J., Shaw, J., and McGrath, M., Eds., Elsevier, New York, 1984, 33.
28. Ross, R. and Kariya, B., Morphogenesis of vascular smooth muscle in atherosclerosis and cell culture, in *Handbook of Physiology - Section 2, The Cardiovascular System, Vol. 2, Vascular Smooth Muscle*, Bohr, D., Somlyo, A. P., and Sparks, H. V., Eds., American Physiological Society, Bethesda, Md., 1980, 69.
29. Chamley, J. H. and Campbell, G. R., Mitosis of contractile smooth muscle cells in tissue culture, *Exp. Cell Res.*, 84, 105, 1974.
30. Mark, G. E., Chamley, J. H., and Burnstock, G., Interactions between autonomic nerves and smooth and cardiac muscle cells in tissue culture, *Dev. Biol.*, 32, 194, 1973.

31. Gröschel-Stewart, U., Chamley, J. H., Campbell, G. R., and Burnstock, G., Changes in myosin distribution in dedifferentiating and redifferentiating smooth muscle cells in tissue culture, *Cell Tissue Res.*, 165, 13, 1975.
32. Purves, R. D., Mark, G. E., and Burnstock, G., The electrical activity of single isolated smooth muscle cells, *Pfluegers Arch.*, 341, 325, 1973.
33. Purves, R. D., Hill, C. E., Chamley, J. H., Mark, G. E., Fry, D. M., and Burnstock, G., Functional autonomic neuromuscular junctions in tissue culture, *Pfluegers Arch.*, 350, 1, 1974.
34. Campbell, G. R., Uehara, Y., Mark, G., and Burnstock, G., Fine structure of smooth muscle cells grown in tissue culture, *J. Cell Biol.*, 49, 21, 1971.
35. Chamley-Campbell, J. H. and Campbell, G. R., Development of smooth muscle and autonomic nerves in culture, in *Somatic and Autonomic Nerve-Muscle Interactions*, Burnstock, G., O'Brien, R., and Vrbova, G., Eds., Elsevier, Amsterdam, 1983, 35.
36. Hermsmeyer, K., De Cino, P., and White, R., Spontaneous contractions of dispersed vascular muscle in cell culture, *In Vitro*, 12, 628, 1976.
37. Mauger, J. P., Worcel, M., Tassin, J., and Courtois, Y., Contractility of smooth muscle cells of rabbit aorta in tissue culture, *Nature (London)*, 255, 337, 1975.
38. Chamley-Campbell, J. H., Nestel, P., and Campbell, G. R., Smooth muscle metabolic reactivity in atherogenesis: LDL metabolism and response to serum mitogens differ according to phenotype, in *Factors in Formation and Regression of the Atherosclerotic Plaque*, Born, G. R. V., Catapano, A. L., and Paoletti, R., Eds., Plenum Press, New York, 1982, 115.
39. Nilsson, J., Sjölund, M., Palmberg, L., Thyberg, J., and Heldin, C.-H., Arterial smooth muscle cells in primary culture produce a platelet-derived growth factor-like protein, *Proc. Natl. Acad. Sci. U.S.A.*, 82, 000, 1985.
40. Campbell, J. H., Popadynec, L., Nestel, P. J., and Campbell, G. R., Lipid accumulation in arterial smooth muscle cells. Influence of phenotype, *Atherosclerosis*, 47, 279, 1983.
41. Campbell, J. H., Reardon, M. F., Campbell, G. R., and Nestel, P. J., Metabolism of atherogenic lipoproteins by smooth muscle cells of different phenotype in culture, *Arteriosclerosis*, 5, 318, 1985.
42. Melzig, von M. and Teuscher, E., The influence of blood factors on the myosin content of vascular smooth muscle cells in culture, *Acta Histochem.*, 73, 289, 1983 (in German).
43. Nilsson, J., Ksiazek, T., Thyberg, J., and Wasteson, A., Cell surface components and growth regulation in cultivated arterial smooth muscle cells, *J. Cell Sci.*, 64, 107, 1983.
44. Busch, C., Ljungman, C., Heldin, C.-M., Waskson, E., and Obrink, B., Surface properties of cultured endothelial cells, *Haemostasis*, 8, 142, 1979.
45. Gamse, G., Fromme, H. G., and Kresse, H., Metabolism of sulfated glycosaminoglycans in cultured endothelial cells and smooth muscle cells from bovine aorta, *Biochim. Biophys. Acta*, 544, 514, 1978.
46. Oohira, A., Wight, T. N., and Bornstein, P., Sulfated proteoglycans synthesized by vascular endothelial cells in culture, *J. Biol. Chem.*, 258, 2014, 1983.
47. Wasteson, Å., Glimelius, B., Busch, C., Westermark, H., Heldin, C.-H., and Norling, B., Effects of a platelet endoglycosidase on cell surface associated heparan sulphate of human cultured endothelial and glial cells, *Thromb. Res.*, 11, 309, 1977.
48. Merrilees, M. J. and Scott, J., Interaction of aortic endothelial and smooth muscle cells in culture. Effect of glycosaminoglycan levels, *Atherosclerosis*, 39, 147, 1981.
49. Heickendorff, L. and Ledet, T., Glycosaminoglycans of arterial basement membrane-like material from cultured rabbit aortic myomedial cells, *Biochim. Biophys. Acta*, 798, 276, 1984.
50. Yamamoto, H., Kanaide, H., and Nakamura, M., Metabolism of glycosaminoglycans of cultured rat aortic smooth muscle cells altered during subculture, *Br. J. Exp. Pathol.*, 64, 156, 1983.
51. Simms, H. S. and Stillman, N. P., Substances affecting adult tissue in vitro. II. A growth inhibitor in adult tissue, *J. Gen. Physiol.*, 20, 621, 1937.
52. Florentin, R. A., Nam, S. C., Janakidevi, K., Lee, K. T., Reiner, J. M., and Thomas, W. A., Population dynamics of arterial smooth-muscle cells. II. In vivo inhibition of entry into mitosis of swine arterial smooth-muscle cells by aortic tissue extracts, *Arch. Pathol.*, 95, 317, 1973.
53. Eisenstein, R., Harper, E., Kuettner, K. E., Schumacher, B., and Matijevitch, B., Growth regulators in connective tissues. II. Evidence for the presence of several growth inhibitors in aortic extracts, *Paroi Arterielle*, 5, 163, 1979.
54. Gajdusek, C., DiCorleto, P., Ross, R., and Schwartz, S. M., An endothelial cell-derived growth factor, *J. Cell Biol.*, 85, 467, 1980.
55. Castellot, J. J., Jr., Addonizio, M. L., Rosenberg, R., and Karnovksy, M. J., Cultured endothelial cells produce a heparinlike inhibitor of smooth muscle cell growth, *J. Cell Biol.*, 90, 372, 1981.
56. Fass, D. N., Downing, M. R., Meyers, P., Bowie, E. J. W., and Witte, L. D., Cell growth stimulation by normal and von Willebrand porcine platelets and endothelial cells, *Blood*, 52 (Suppl. 1), 181, 1978.
57. Wang, C.-H., Largis, E. E., and Schaffer, S. A., The effects of endothelial cell-conditioned media on the proliferation of aortic smooth muscle cells and 3T3 cells in culture, *Artery*, 9, 358, 1981.

58. Witte, L. D., Cornicelli, J. A., Miller, R. W., and Goodman, D. S., Effects of platelet-derived and endothelial cell-derived growth factors on the low density lipoprotein receptor pathway in cultured human fibroblasts, *J. Biol. Chem.*, 257, 5392, 1982.
59. Bowen-Pope, D. F., Vogel, A., and Ross, R., Production of platelet-derived growth factor-like molecules and reduced expression of platelet-derived growth factor receptors accompany transformation by a wide spectrum of agents, *Proc. Natl. Acad. Sci. U.S.A.*, 81, 2396, 1984.
60. DiCorleto, P. E., Cultured endothelial cells produce multiple growth factors for connective tissue cells, *Exp. Cell Res.*, 153, 167, 1984.
61. DiCorleto, P. E. and Bowen-Pope, D. F., Cultured endothelial cells produce a platelet-derived growth factor-like protein, *Proc. Natl. Acad. Sci. U.S.A.*, 80, 1919, 1983.
62. Castellot, J. J., Jr., Favreau, L. V., Karnovsky, M. J., and Rosenberg, R. D., Inhibition of vascular smooth muscle cell growth by endothelial cell-derived heparin. Possible role of a platelet endoglycosidase, *J. Biol. Chem.*, 257, 11256, 1982.
63. Hatcher, V. B., Tsien, G., Oberman, M. S., and Burk, P. G., Inhibition of cell proliferation and protease activity by cartilage factors and heparin, *J. Supramol. Struct.*, 14, 33, 1980.
64. Hoover, R. L., Rosenberg, R., Haering, W., and Karnovsky, M. J., Inhibition of rat arterial smooth muscle cell proliferation by heparin. II. In vitro studies, *Circ. Res.*, 47, 578, 1980.
65. Fritze, M. S., Reilly, C. F., and Rosenberg, R. D., An antiproliferative heparan sulfate species produced by postconfluent smooth muscle cells, *J. Cell Biol.*, 100, 1041, 1985.
66. Kjellén, L., Pettersson, I., and Höök, M., Cell-surface heparan sulfate: an intercalated membrane proteoglycan, *Proc. Natl. Acad. Sci. U.S.A.*, 78, 5371, 1981.
67. Palotie, A., Tryggvason, K., Peltonen, L., and Seppä, H., Components of subendothelial aorta basement membrane. Immunohistochemical localization and role in cell attachment, *Lab. Invest.*, 49, 362, 1983.
68. Wight, T. N., Vessel proteoglycans and thrombogenesis, in *Progress in Hemostasis and Thrombosis*, Spaet, T., Ed., Grune & Stratton, New York, 1980, 1.
69. Campbell, G. R. and Campbell, J. H., Smooth muscle phenotypic changes in arterial wall homeostasis: implications for the pathogenesis of atherosclerosis, *Exp. Mol. Pathol.*, 42, 139, 1985.
70. Rosenberg, R. D., Castellot, J. J., Jr., and Karnovsky, M. J., Role of heparin-like molecules in atherogenesis, in *Heparin: New Biochemical and Medical Aspects*, Witte, I., Ed., Walter de Gruyter, Berlin, 1982, 261.
71. Lash, J. W. and Vasan, N. S., Somite chondrogenesis in vitro. Stimulation by exogenous extracellular matrix components, *Dev. Biol.*, 66, 151, 1978.
72. Hay, E. D., Collagen and embryonic development, in *Cell Biology of Extracellular Matrix*, Hay, E. D., Ed., Plenum Press, 1981, 379.
73. Archer, C. W., Rooney, P., and Wolpert, L., Cell shape and cartilage differentiation of early chick limb bud cells in culture, *Cell Differ.*, 11, 245, 1982.
74. Geer, J. C. and Haust, M. D., Smooth muscle cells in atherosclerosis, in *Monographs on Atherosclerosis*, Vol. 2, Pollak, O. J., Simms, H. S., and Kirk, S. E., Eds., S. Karger, Basel, 1972.
75. McGill, H. C., Jr., The lesion, in *Atherosclerosis. III. Proc. 3rd Int. Symp. Atherosclerosis*, Schettler, G. and Weizel, A., Eds., Springer-Verlag, Berlin, 1974, 27.
76. Campbell, G. R. and Chamley-Campbell, J. H., The cellular pathobiology of atherosclerosis, *Pathology*, 13, 423, 1981.
77. Velican, C. and Velican, D., Intimal thickening in developing coronary arteries and its relevance to atherosclerotic involvement, *Atherosclerosis*, 23, 345, 1976.
78. Velican, C. and Velican, D., The precursors of coronary atherosclerotic plaques in subjects up to 40 years old, *Atherosclerosis*, 37, 33, 1980.
79. Haust, M. D., Atherosclerosis and smooth muscle cells, in *Biochemistry of Smooth Muscle*, Stephens, N. L., Ed., CRC Press, Boca Raton, Fla., 1983, 189.
80. Orekhov, A. N., Karpova, I. I., Tertov, V. V., Rudchenko, S. A., Andreeva, E. R., Krushinsky, A. V., and Smirnov, V. N., Cellular composition of atherosclerotic and uninvolved human aortic subendothelial intima. Light-microscopic study of dissociated aortic cells, *Am. J. Pathol.*, 115, 17, 1984.
81. Campbell, G. R. and Campbell, J. H., Smooth muscle cells, in *The Biology and Clinical Science of Atherosclerosis*, European Atherosclerosis Group, Churchill Livingstone, Edinburgh, 1986, in press.
82. Mosse, P. R. L., Campbell, G. R., Wang, T. L., and Campbell, J. H., Smooth muscle phenotypic expression in human carotid arteries. I. Comparison of cells from diffuse intimal thickenings adjacent to atheromatous plaques with those of the media, *Lab. Invest.*, 53, 556, 1985.
83. Mosse, P. R. L., Campbell, G. R., and Campbell, J. H., Smooth muscle phenotypic expression in human carotid arteries. II. Comparison of cells from areas of atherosclerosis-free intimal thickenings with those of the media, *Arteriosclerosis*, 6(6), 1986.
84. Gabbiani, G., Kocher, O., Bloom, W. S., Vandekerckhove, J., and Weber, K., Actin expression in smooth muscle cells of rat aortic intimal thickening, human atheromatous plaque, and cultured rat aortic media, *J. Clin. Invest.*, 73, 148, 1984.

85. Ross, R., Wight, T. N., Strandness, E., and Thiele, B., Human atherosclerosis. I. Cell constitution and characteristics of advanced lesions of the superficial femoral artery, *Am. J. Pathol.*, 114, 79, 1984.

Chapter 5

CONTRACTILE PROTEIN EXPRESSION AND CYTODIFFERENTIATION IN CULTURED VASCULAR SMOOTH MUSCLE CELLS

Gary K. Owens, Arthur S. Rovner, and Richard A. Murphy

TABLE OF CONTENTS

I. INTRODUCTION

There is much interest in the development of cultured vascular smooth muscle (VSM) cells for use in studying receptor or contractile function in vitro. Isolated cells have unique advantages for studying mechanisms of action of vasoactive substances without the complicating influences of other cell types, diffusional limitations, innervation, as well as technical limitations in the types of measurements that can be performed on a whole tissue (e.g., optical measurements of Ca^{2+} concentrations using fluorescent indicators). However, Chamley-Campbell et al.[1] and others[2] presented electron microscopic and immunocytochemical evidence suggesting that VSM cells do not proliferate in cell culture until they have lost much of their contractile apparatus. The Campbells suggested that VSM cells normally exist in a nondividing "contractile state" and phenotypically modulate to a "synthetic state" in cell culture. Other studies, however, have demonstrated that established cultures of VSM cells express muscle-specific contractile proteins[3-8] and exhibit some contractile properties[9-12] characteristic of differentiated smooth muscle.

In this chapter we will

1. Review the composition, structure, and function of the contractile apparatus in smooth muscle cells
2. Present evidence showing that VSM cells continue to express smooth muscle-specific contractile proteins in culture and that this expression is related to the cell's growth state
3. Examine whether additional factors characterizing VSM differentiation are also promoted by conditions that increase expression of smooth muscle cell-specific contractile proteins
4. Review known factors that promote cytodifferentiation in cultured VSM cells

II. CONTRACTILE SYSTEM OF SMOOTH MUSCLE CELLS IN VIVO

Smooth muscle, like striated muscle, is a specialized cell type with an organized myofilament lattice for the expression of external force or shortening. There is a generally shared consensus, based primarily on mechanical and biochemical analysis, that contraction is due to a sliding filament/crossbridge mechanism. Interdigitation of thin filaments, consisting of actin and tropomyosin, into a myosin-containing thick filament lattice appears to reflect crossbridge cycling powered by ATP hydrolysis.[13-15] However, the assumption that smooth and striated muscles differ significantly only in the structural organization of the contractile proteins is incorrect. This section summarizes recent information showing that the contractile system of smooth muscle has unique properties underlying the ability to perform a variety of functions.

A. Characteristics of the Contractile System

The ultrastructural appearance of smooth muscles, typified by a densely packed array of axially oriented thin filaments with scattered thick filaments, is paralleled by a correspondingly high content of thin filament proteins and low myosin content compared to mammalian skeletal muscles (Table 1). Within the precision of the estimates, all mammalian smooth muscles have a comparable myosin content, but arteries differ from veins and a wide variety of other tissues in having cells with higher thin filament protein contents.

The estimated mechanical output of smooth muscle cells also contrasts with striated muscle (Table 1). These figures are based on data obtained from the few preparations

Table 1
CONTRACTILE PROTEIN COMPOSITION AND MECHANICAL OUTPUT

	Content[a] (mg/g cell wet weight)			Mechanical output	
	Myosin	Actin	Tropomyosin	Isometric stress[b] ($\times 10^5$ N/m^2)	V_0[c] (L_0/sec)
Skeletal muscle	62	22	5	2—3.5	4—24
Arterial smooth muscle	18	50	13	Up to 6.7	0.1
Veins and smooth muscles of airway, gastrointestinal, and other organs	18	29	8	Up to 7.3	0.4

[a] Summary from Murphy et al.[16]

[b] Maximum reported estimates for cells (may be underestimates for smooth muscles due to tissue damage, incomplete activation): swine carotid media,[17] guinea pig taenia coli.[18]

[c] Values for mammalian skeletal muscles depend on myosin isoenzyme.[19] Values for smooth muscle also depend on level of crossbridge phosphorylation. Values for maximal unloaded shortening velocities were extrapolated for 100% phosphorylation: swine carotid media,[16] feline esophageal,[20] bovine trachealis.[21]

in which tissue measurements can be corrected to provide data on cellular output and in which efforts have been made to achieve maximal activation. Two striking differences distinguish smooth and striated muscles. The latter have very high estimated unloaded shortening velocities (V_0). Differences in V_0 among striated muscle cells reflect synthesis of different myosin isoenzymes. The velocity of a given striated muscle cell also depends on load, but V_0 is not directly determined by cell [Ca^{2+}]. In contrast, smooth muscles are slow, and the V_0 for a given cell can vary over a wide range.[22] In smooth muscle, crossbridge cycling rates appear to be a regulated parameter reflecting cell [Ca^{2+}] and the level of crossbridge phosphorylation. Thus, smooth muscles have properties that are absent in striated muscles.

Smooth muscles also differ from striated muscles by generating higher levels of isometric stress (force/cross-sectional area of cells). This difference represents a major puzzle because the myosin and crossbridge content of smooth muscles is only 30% of that in striated muscles (Table 1). Stresses and velocities in muscles depend not only on the nature of the contractile proteins, but also on their organization into the filament lattice. The latter is not known in smooth muscles, and the relative contributions of protein differences and filament lattice structure to estimated cellular contractile outputs is uncertain.

B. Contractile Function in Smooth Muscle

Skeletal muscle cells extend from one tendon to another. The skeleton bears most gravitational loads, and the cells are normally relaxed, contracting when needed to produce work. Furthermore, each cell (or motor unit) is recruited and functions independently. These conditions do not apply to smooth muscle cells in the walls of hollow organs. The connection of the small cells in series around the organ requires that such cells must generate and bear equal stresses. Furthermore, contraction in part of a tissue produces changes in pressure that are transmitted to all other cells, so the elements are not independent. A final point is that smooth muscles function not only to produce work, but must be capable of tonic contractions to maintain the circumference or volume of the organ (e.g., vascular resistance). Smooth muscles exhibit a wide variety of activity patterns ranging from normal tonic contractions (i.e., sphincters) to intermittent phasic contractions comparable to skeletal muscle twitches. The available evidence is consistent with the hypothesis that all smooth muscles have a similar contractile apparatus and that unique Ca^{2+}-dependent regulatory mechanisms allow very different types of contractions.[22] Physiological diversity appears to arise from differences in the ways tissues control myoplasmic [Ca^{2+}].

The flexibility of the smooth muscle contractile apparatus, at least in part, arises from properties which allow two types of crossbridge interactions with the thin filaments.[22-25] Contraction of a relaxed cell is initiated by a rapid increase in myoplasmic [Ca^{2+}] to levels that activate myosin kinase. This leads to phosphorylation of the crossbridges at a site on the regulatory light chains of myosin. Phosphorylated crossbridges cycle rapidly and produce fast responses, with correspondingly high rates of energy consumption. However, activation is characteristically associated with a transient increase in myoplasmic [Ca^{2+}], which falls to an intermediate sustained level.[26] Mechanical studies indicate that reductions in cell [Ca^{2+}] to intermediate levels leading to dephosphorylation by myosin phosphatase are associated with the formation of a non- or very slowly cycling attachment of the crossbridges with the thin filament.[23,24] In this state, the contractile system can maintain a stress against an imposed load (such as blood pressure) with greatly reduced cycling and a very high economy. This state has been termed "latch" and appears to characterize tonic contractions in sphincters[20] or arterial smooth muscle.[23,24] The latch state is very sensitive to Ca^{2+}, but reductions in Ca^{2+} lead to relaxation.[27] In summary, any analysis of the contractile protein variants expressed in smooth muscle must consider their role in the function of the tissue, which has elements not exhibited by striated muscle.

C. Molecular Diversity of the Contractile Proteins and Functional Significance

Striated muscles express a single actin variant. The functional differences between skeletal muscle cell types are associated with isoenzymes of myosin and the metabolic adaptations necessary to match ATP production with consumption in cells with widely differing shortening velocities. There is no compelling evidence for the expression of myosin variants in vertebrate smooth muscle[16] except for an avian embryonic form.[28] Several laboratories have reported preliminary evidence that smooth muscles have two or more myosin heavy chains differing in their molecular weights.[29-31] However, proteolytic cleavage during tissue processing readily occurs, leading to possible artifacts. Our studies[31] show equal amounts of two heavy chains in several types of smooth muscle. However, only a single myosin type was detected on nondenaturing pyrophosphate gels. These observations are consistent with an interpretation that mature smooth muscle contains a single myosin form that is a heterodimer consisting of two different heavy chains (plus two 20,000- and two 17,000-dalton light chains). If this hypothesis is correct, the two heads of the crossbridge in smooth muscle may differ, with possible functional significance with respect to crossbridge interactions with the thin filament discussed above.

Smooth muscle myosin is immunologically and electrophoretically distinct from the isomyosins of striated muscles and that present in nonmuscle cells (Figure 1).[32-35] Although myosin variants have not been shown to contribute to functional diversity, the expression of the smooth muscle-specific myosin or its denatured heavy chains is a readily measured and potentially useful marker for assessing cytodifferentiation in cultured or developing smooth muscle.

In contrast to striated muscle, the molecular diversity in the contractile proteins of smooth muscle is found in the thin filament proteins. Vandekerckhove and Weber[38] demonstrated that four forms of actin are expressed in smooth muscles. These include two smooth muscle-specific actins (α-SM and γ-SM) and two variants expressed in nonmuscle cells (β-NM and γ-NM), where the α, β, and γ designate variants distinguish-

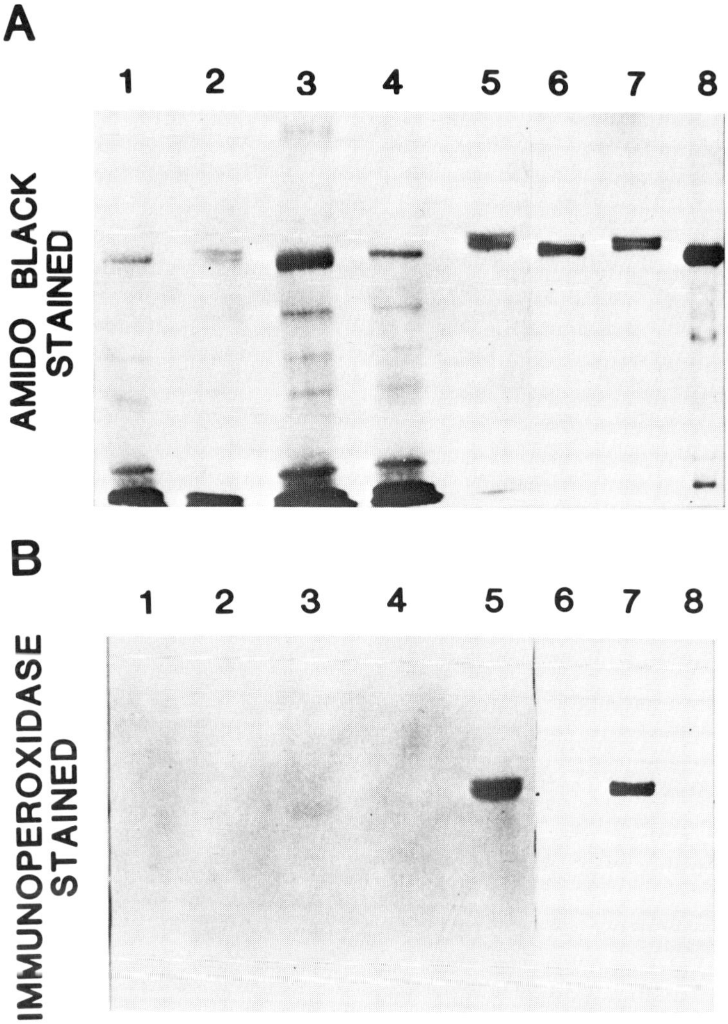

FIGURE 1. Immunological characterization of myosins from a variety of cells and tissues, using the smooth muscle myosin-specific antibody of Groschel-Stewart et al.[37] Myosins were resolved by sodium dodecyl sulfate (SDS) gel electrophoresis (3 to 4% gradient gel) and blotted onto nitrocellulose paper.[38] Panel A shows an amido black stain illustrating the proteins present on the nitrocellulose. Panel B shows results of immunoperoxidase staining of these same gels with the Groschel-Stewart et al.[37] anti-myosin antibody. Lanes: (1) Swiss 3T3 cells; (2) cultured rat coronary artery endothelial cells; (3) cultured rat aortic smooth muscle cells at confluency; (4) cultured rat aortic smooth muscle cells during log phase growth; (5) purified myosin from swine stomach; (6) purified rabbit skeletal muscle myosin; (7) purified myosin from the media of swine carotid artery; and (8) rat left ventricle tissue homogenate. Note that myosin was present in all lanes (Panel A), but only those from swine stomach (Lane 5) and carotid artery (Lane 7) reacted with the Groschel-Stewart antimyosin antibody. In contrast, the myosins from all cultured cells reacted with an antiplatelet myosin antibody provided by D. Larsen (data not shown). Controls were done using nonimmune serum (data not shown).

able on the basis of their isoelectric points. Only three variants are found in two-dimensional gels of smooth muscle proteins because the γ-SM and γ-NM actin variants comigrate. Despite this limitation, two-dimensional gel electrophoresis provides a relatively simple method to estimate actin variant contents in smooth muscle tissue samples lacking significant amounts of other cell types.[39] Our results[6,7,39] and those of others[40,41] clearly show that the actin variants are highly tissue-specific. All arteries, for example, exhibit comparable actin variant distributions irrespective of species. Measurements of the isoactin distribution and content has proven to be a useful and readily determined marker for studies of cytodifferentiation in smooth muscle.

Expression of muscle (i.e., α-SM and γ-SM) and nonmuscle (β-NM and γ-NM) actins appears to be developmentally regulated in smooth muscle. Saborio et al.[42] found a continuous increase in γ-SM actin and a decrease in nonmuscle actins in the chicken gizzard during embryonic development and showed that the differential expression of actins was controlled at the transcription level. We demonstrated that similar changes occur during developmental growth of VSM cells.[6] β-NM actin comprised about 60% of the total actin in VSM cells from newborn rats. The amount of β-NM actin decreased, whereas the α-SM actin increased with age, comprising 70% of the actin in VSM cells from adult rats.

The functional significance of changes in isoactin expression during developmental growth of VSM is unclear, but may reflect the physiological requirements for specific actin variants for specialized cellular functions. There is considerable evidence supporting this idea in skeletal muscle. Whereas prefusion skeletal myoblasts contain principally nonmuscle isoactins, a dramatic change in isoactin expression occurs after fusion, with α-actin (the skeletal muscle type) becoming the only significant actin variant present in differentiated skeletal muscle.[43,44] Concurrent with these changes in actin expression, extensive changes in the structure and function (i.e., sarcomeres form and cells become contractile) of actin occur in these cells. Recent immunocytological studies by Pardo et al.,[45] demonstrating differential subcellular localization of muscle and nonmuscle actins in skeletal muscle cells, also support the idea that isoactins have specialized roles within the cell. However, all actins are similar in their primary structure,[38] physical properties,[45] and actin binding sites.[47] Therefore, the precise nature of purported functional differences between isoactins is unclear. Despite these uncertainties, measurements of changes in isoactin expression appear to be a very useful marker for studying cytodifferentiation of smooth muscle during development or culture.

III. CONTRACTILE PROTEIN EXPRESSION IN CULTURED VASCULAR SMOOTH MUSCLE CELLS

A. Actin Expression in Cultured Vascular Smooth Muscle Cells

Our laboratory has been engaged in developing a functional culture model for studying VSM contractile responsiveness in vitro. However, assessment of contractile responsiveness is difficult to measure quantitatively in cultured cells. Therefore, our research strategy is to study factors that promote cytodifferentiation in cultured smooth muscle cells, utilizing expression of the smooth muscle-specific forms of actin (i.e., α-SM and γ-SM isoactins) as a marker for differentiation in these cells. Changes in isoactin expression in relation to growth state were evaluated in both primary and subpassaged rat aortic smooth muscle cells obtained by enzyme dispersion techniques.[1,48] Isoactin expression was evaluated by (1) measurements of relative isoactin content and rates of synthesis (^{35}S-methionine incorporation) by densitometric evaluation[49] of two-dimensional isoelectric focusing (IEF)-sodium dodecyl sulfate (SDS) gels[50] and (2) by immunocytological examination of individual smooth muscle cells utilizing a smooth muscle-specific isoactin antibody (designated "CGA7"),[51] provided by Gown and Gordon (University of Washington, Seattle). These studies[6,7] yielded the following results:

1. Gel electrophoretic analysis showed no detectable decreases in α-smooth muscle isoactin content or myosin content (heavy chain) in primary cultures of VSM cells prior to the onset of cellular proliferation. In addition, cells identified as undergoing DNA synthesis by ^{3}H-thymidine autoradiography simultaneously stained with the smooth muscle-specific isoactin antibody. Thus, loss of smooth muscle-specific isoactins and myosin was not a prerequisite for initiation of cellular proliferation. Significant levels of α-SM actin were present at the onset of proliferation, although the fractional synthesis of this actin had virtually ceased (Figure 2).
2. Expression of smooth muscle-specific isoactins (synthesis and content) was drastically reduced in log phase growth cells and then increased in association with density-induced decreases in cellular growth rates as cells reached confluency (Figure 2). Conversely, β-NM actin synthesis and content were increased in rapidly dividing smooth muscle cells, but decreased in density-arrested cells. Significantly, these changes were observed in both primary and subpassaged VSM cultures.
3. There was a dramatic decrease (Figure 3a) in the relative synthesis of α-smooth muscle isoactin when saturation density cells were passaged and plated at low densities in serum-containing medium, which persisted until cells reached confluency (Figure 3d). However, this decrease could be almost totally prevented by plating cells in serum-free medium[52] (data not shown), plasma derived serum (Figure 3c),[53] or by plating cells in serum at a density exceeding saturation density (Figure 3b). If cells were initially plated at a sparse density in the presence of serum and then switched to serum-free conditions, only moderate increases in the relative synthesis of α-SM isoactin occurred within the first 24 hr.
4. Serum stimulation of cells that had been growth-arrested in either plasma-derived serum[53] or serum-free medium[52] resulted in a dramatic decrease in α-SM actin expression and an increase in β-NM expression that preceded the onset of cellular proliferation.[7]

These findings demonstrate that cultured VSM cells undergo differential expression of isoactins in relation to their growth state, and suggest that growth arrest of cells promotes cytodifferentiation in cultured VSM cells. Strauch and Rubenstein[5] reported similar findings in studies of BC_3H_1 cells, a smooth muscle-like cell line derived from a mouse brain tumor. Similarly, Franke et al.[3] demonstrated continued expression of the m-RNA for α-SM actin in an established smooth muscle cell line derived from the rat inferior vena cava. Larsen et al.[8] described continued myosin expression in primary and established cultures of rat mesenteric VSM cells. Thus, loss of cell-specific contractile proteins does not appear to be prerequisite for cellular proliferation in cultures of VSM cells. However, our studies[6,7] and those of Strauch and Rubenstein[5] do show that smooth muscle cells in both primary and subpassage undergo differential expression of isoactins in relation to their growth state. We want to emphasize that these results provide evidence that expression of cell-specific characteristics and cellular growth are coupled, rather than evidence that phenotypic modulation is a prerequisite for smooth muscle cell proliferation.

It is important to note that these studies also suggest that cell-cell contact has an important permissive effect on induction of α-SM actin synthesis. Strauch and Rubenstein[5] found that α-SM actin synthesis could be induced much sooner in BC_3H_1 cells by growth-arresting confluent cells by withdrawal of serum, but no changes were ob-

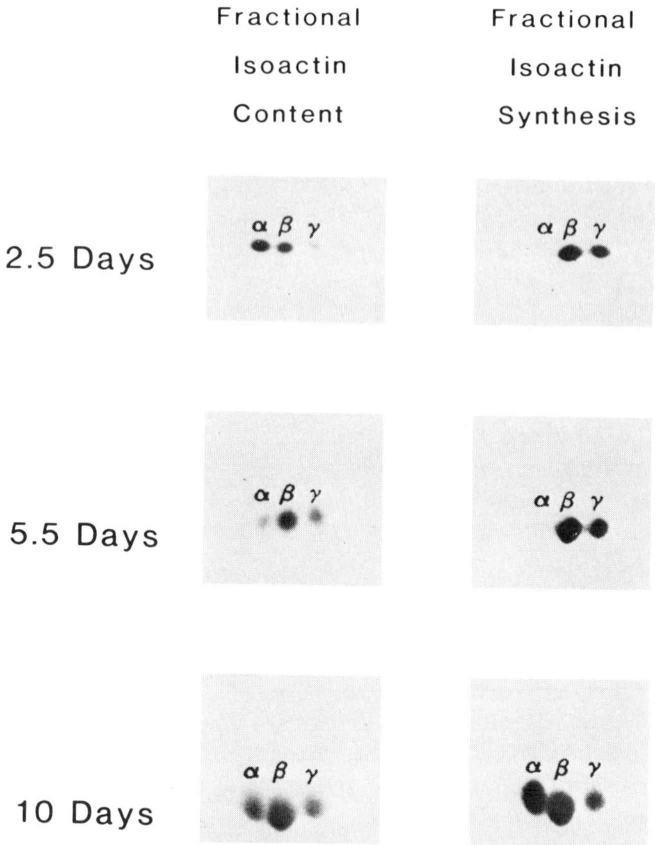

FIGURE 2. Two-dimensional isoelectric focusing — SDS gels showing the fractional content (silver-stained gel) and synthesis (^{35}S-methionine autoradiographs) of isoactins in rat aortic smooth muscle cells at 2.5 days, 5.5 days, and 10.0 days in primary culture. For measurements of isoactin synthesis, duplicate culture dishes were pulsed with ^{35}S-methionine (4 hr, 80 μCi/mℓ, 986 to 1103 Ci/mmol). Samples were harvested, pooled, solubilized, and isoactins resolved by two-dimensional gel electrophoresis.[50] Replicate dishes were pulsed with ^{5}H-thymidine (1 μCi/mℓ, 1 hr) for determination of the frequency of replicating cells (data not shown). At 2.5 days a large fraction of cells were labeled with ^{3}H-thymidine, and α-SM actin content was high, although α-SM actin synthesis was very low. Cells were undergoing logarithmic growth at 5.5 days, and both α-SM actin content and synthesis were extremely low. Cells were postconfluent at 10.0 days; growth rates had decreased dramatically, and both α-SM actin content and synthesis had increased substantially. α = α-SM actin; β = β-NM actin; γ = γ-SM + γ-NM actin. (Reproduced from *The Journal of Cell Biology*, 1986, Vol. 102, pp. 343-352 by copyright permission of the Rockefeller University Press.)

served if cells were serum-deprived when subconfluent. In our studies, induction of α-SM isoactin coincided with attainment of cell confluency, although maximal α-SM isoactin expression did not occur until cells were several days postconfluent. Whereas high levels of isoactin synthesis could be maintained in sparse cultures in serum-free medium or plasma-derived serum, we found that growth arrest of sparse density cells that were not already induced was not very effective in increasing α-SM actin expression unless serum withdrawal was for extended time periods. Furthermore, the fact that cells continue to express high levels of nonmuscle actins even when growth-ar-

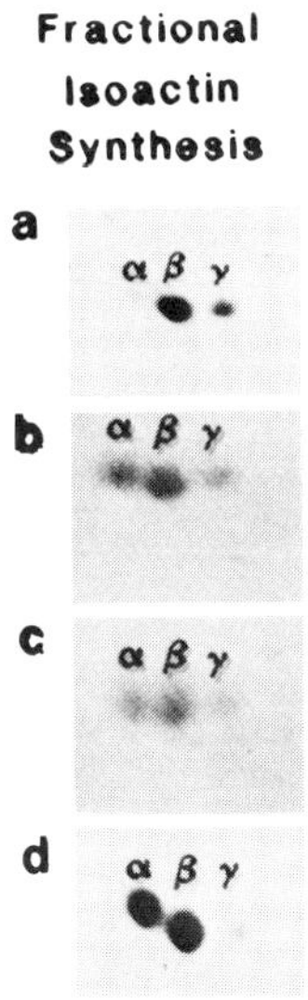

FIGURE 3. ^{35}S-methionine autoradiographs of two-dimensional IEF/SDS gels illustrating the effects of cellular growth state on the fractional synthesis (i.e., ^{35}S-methionine autoradiographs) of isoactins in subpassaged rat aortic smooth muscle cells. Cells at saturation density were harvested with a trypsin-EDTA solution and plated in the medium indicated. All cells other than those illustrated in (b) were plated at 10^4 per square centimeter. Replicate dishes of each group were pulsed for 1 hr with ^{3}H-thymidine for determination of the frequency of cells undergoing DNA replication (data not shown). (a) 2.5 days after plating in 10% fetal calf serum (FCS) (cells were growing logarithmically); (b) 2.5 days after plating in 10% FCS at a density (i.e., 5.0×10^5 cm^{-2}) exceeding saturation density for these cells (replication rates were <1%); (c) 2.5 days following plating in plasma-derived serum[53] (the frequency of replicating cells was <5%); (d) 10 days after plating in 10% FCS (cells were postconfluent, and the frequency of replicating cells was <2%). Note that the fractional synthesis of α-SM actin is markedly increased in growth-arrested cells (b to d) as compared to rapidly dividing cells (a). α = α-SM actin; β = β-NM actin; γ = γ-smooth muscle actin + γ-NM actin.

rested indicates that additional criteria other than cessation of an initiation of cellular proliferation must be important in inducing cytodifferentiation in these cells.

B. Myosin Expression in Cultured Smooth Muscle Cells

Using electron microscopy and a smooth muscle cell-specific myosin antibody,[37] Chamley-Campbell and co-workers[1,55] reported that VSM cells characteristically lose myosin prior to the onset of cellular proliferation and only variably regain it at confluency, depending on culture conditions and initial plating densities. However, SDS-polyacrylamide gel electrophoresis shows that smooth muscle myosin is present at all times in primary and passaged smooth muscle cell cultures (see Figure 1).[7,8] Larsen et al.[8] found changes in VSM myosin antigenic expression in culture, but presented convincing evidence that the changes reflected conformational alterations rather than synthesis of a new myosin variant. Recent studies by Rovner et al.[56] confirm the results of Owens et al.[7] and Larsen et al.[8] and appear to resolve the apparent discrepancies between the findings of these workers and those of Chamley-Campbell. Based on Western blot analysis, Rovner et al.[56] found that the antisera utilized by Chamley-Campbell et al.[1] is specific for smooth muscle myosin and does not stain myosin isolated from a variety of nonmuscle cells or tissues or from skeletal and cardiac muscles (Figure 1). In agreement with observations of Larsen et al.[8] and Owens et al.,[7] some form of myosin was present in cultured smooth muscle cells at all times by SDS-PAGE. However, the myosin antibody utilized by Chamley-Campbell did not react with the myosin derived from cultured VSM cells in Western blot experiments, but the antiplatelet myosin (APM) and antismooth muscle myosin (ASMM) (both provided by Larsen) did react (data not shown). Our interpretation of these results is that VSM cells synthesize a myosin in vitro that is distinct from the variant found in vivo, and that the conformational changes observed by Larsen et al.[8] are not necessarily related to this phenomenon. The studies of Rovner et al.[56] also demonstrated that the myosin heavy chain present in cultured cells consisted of a single molecular weight variant of approximately 205,000 daltons, while the myosin expressed in vivo appears to be a heterodimer consisting of two heavy chains of approximately 207,000 and 211,000 daltons.[31] The molecular weight of the myosin heavy chain derived from cultured VSM cells was identical to that found in several nonmuscle cells. Studies are in progress to determine under what conditions the in vivo myosin variants might be expressed in cultured VSM cells. It is interesting that Chamley-Campbell et al.[1,55] observed that cultured VSM cells characteristically lose smooth muscle-specific myosin staining at the onset of cell proliferation, but regain it at confluency if plated at a high initial density.[55] In contrast, neither cells plated at low densities nor subpassaged cells showed positive staining with this antibody under their culture conditions.

C. Contractile Responsiveness of Cultured Vascular Smooth Muscle Cells

Cultured VSM cells do not generally appear to be capable of force generation. However, there is now good evidence showing that several elements of excitation-contraction coupling are functional in cultured cells. Established cultures of rat mesenteric smooth muscle cells retain high affinity receptors to angiotensin II and undergo dose-dependent A-II-induced phosphorylation of the 20,000-dalton myosin light chain.[57] Furthermore, a variable but significant proportion of these cells contracted (reversible, agonist-specific cell shortening) in response to angiotensin II.[9] Martin and Gordon[10] observed changes in ^{86}Rb efflux in cultured rabbit aortic smooth muscle cells in response to stimulation with norepinephrine or histamine. In addition, Colucci et al.[12] recently demonstrated α_1-adrenergic receptor-coupled calcium flux in cultured smooth muscle cells derived from rabbit aorta. Thus, it appears that important steps in receptor activation-contraction coupling continue to be expressed in vitro.

Cultured VSM cells are, however, "phenotypically modulated"* according to many criteria.[1,2,58] For example, they have a decreased actin thin filament content, increased synthetic cellular organelles, and lack ultrastructurally demonstrable myosin-containing thick filaments, although cultured cells do have prominent cytoplasmic stress fibers. Consistent with the morphological data, cultured cells have a decreased absolute content of actin and myosin,[7,8] exhibit increased proliferative and biosynthetic activities,[1,48] and with few exceptions, do not appear to be capable of stress development. However, observations that subpassaged smooth muscle cells continue to express a number of cell-specific contractile proteins (e.g., smooth muscle-specific isoactins, myosin light chain kinase) and that cells continue to show a variety of characteristics indicative of contractile responsiveness, indicate that the concept of a "synthetic state" vs. "contractile state" is somewhat misleading and overly simplistic in describing the differentiated state of VSM cells in culture.**

* Chamley-Campbell et al.[1] have utilized the term "modulated" to refer to changes in phenotype from a predominantly contractile state to a predominantly synthetic state. This term is more appropriate than the often-used term "dedifferentiated", which is clearly incorrect since there is no evidence that VSM cells return to a multipotent state.

** While the usage of the terms "synthetic state" and "contractile state" are not technically incorrect, they imply absolute rather than relative changes in cell phenotype. The usage of these terms in the literature has often not been that originally intended by Chamley-Campbell et al.[1]

IV. FACTORS CONTROLLING VSM CELL CYTODIFFERENTIATION IN CULTURE

Our understanding of the factors influencing the expression of cytodifferentiated characteristics in cultured VSM cells is extremely limited. As noted previously, our studies[6,7] and those of Strauch and Rubenstein[5] demonstrated the importance of cell density and growth state in controlling expression of smooth muscle cell-specific isoactins. Furthermore, there is substantial evidence suggesting that other differentiated properties of smooth muscle cells are also promoted in association with growth arrest. Schubert et al.[59] demonstrated an increase in the specific activity of creatine phosphokinase and myokinase associated with increased cell density and the cessation of cell division in BC_3H_1 cells. Subsequently, Olson et al.[60] demonstrated that the induction of creatine phosphokinase in these cells occurred at the transcriptional level. Chamley-Campbell and Campbell[55] presented morphological evidence suggesting that the loss of contractile myofilaments in primary smooth muscle cell culture could be inhibited by plating cells at high density, by growth in medium containing heparin or by growth in co-culture with endothelium (see Chapter 4). Martin and Gordon[10] found that the responsiveness of primary cultures of rabbit aortic smooth muscle cells to agonist-induced changes in membrane conductance could be promoted with heparin. Observations with heparin and endothelium are relevant to this discussion, since Karnovsky and co-workers[61] reported that heparin inhibits smooth muscle cell proliferation in vivo and in vitro, and that endothelial cells produce a heparin-like inhibitor of smooth muscle cell growth (see Chapter 7). However, it is not clear from these cited studies whether the effects of heparin on smooth muscle cell differentiation were in fact related to growth inhibition. Furthermore, the nature of the endothelial-derived differentiation-promoting factor is unclear.

Chamley-Campbell et al.[1] presented evidence suggesting that "phenotypic modulation" in primary smooth muscle cell cultures can be delayed in sparse VSM cell cultures by factors produced by confluent smooth muscle cell cultures, by growth in the presence of sympathetic neurons or nerve extracts, and by growth in medium containing dibutyryl cyclic AMP or theophylline (see Chapter 17). However, these studies were based on morphological evaluations, and the relationship of morphologic changes to smooth muscle cell contractile responsiveness is uncertain. Nevertheless, these results are intriguing, and suggest a number of approaches that should be further examined using more definitive criteria for assessing the differentiated state of VSM cells. Our studies suggest that the expression of smooth muscle cell-specific isoactins and myosin heavy chains will be useful in this regard. However, we want to emphasize that it is important to obtain measures of contractile responsiveness as well.

V. SUMMARY AND CONCLUSIONS

VSM cells undergo marked changes in phenotype when grown in cell culture. However, cells continue to express a variety of cell-specific characteristics that distinguish them as smooth muscle cells. Whereas cultured cells appear to have a much greater synthetic capability than their in vivo counterparts, the concept of a "contractile" vs. a "synthetic" state fails to adequately describe the differentiated state of VSM cells in culture.

There is now good evidence that the expression of cytodifferentiated characteristics in cultured VSM cells is closely related to their growth state and can be promoted by modification of their growth medium. Furthermore, there are now a variety of definitive, quantitative markers of VSM cytodifferentiation that can be systematically applied to studies of VSM cytodifferentiation. In particular, we can now begin to explore the molecular controls of VSM differentiation in a manner similar to that done in skeletal muscle. These studies may result in more useful cultured smooth muscle cells for studying contractile function in vitro. More importantly, such studies will contribute substantially to our understanding of the interrelationship between growth and cytodifferentiation in VSM cells, which, at present, is a poorly understood area of smooth muscle cell biology.

REFERENCES

1. Chamley-Campbell, J., Campbell, G. R., and Ross, R., The smooth muscle cell in culture, *Physiol. Rev.*, 59, 1, 1979.
2. Thyberg, J., Palmberg, L., Nilsson, J., Ksiazek, T., and Sjolund, M., Phenotype modulation in primary cultures of arterial smooth muscle cells on the role of platelet-derived growth factor, *Differentiation*, 25, 156, 1983.
3. Franke, W. W., Schmid, E., Vandekerckhove, J., and Weber, K., A permanently proliferating rat vascular smooth muscle cell with maintained expression of smooth muscle characteristics, including actin of the vascular smooth muscle type, *J. Cell Biol.*, 87, 594, 1980.
4. Gabbiani, G., Kocher, O., Bloom, W. S., Vandekerckhove, J., and Weber, K., Actin expression in smooth muscle cells of rat aortic intimal thickening, human atheromatous plaque, and cultured rat aortic media, *J. Clin. Invest.*, 73, 148, 1984.
5. Strauch, A. R. and Rubenstein, P. A., Induction of vascular smooth muscle α-isoactin expression in BC_3H_1 cells, *J. Biol. Chem.*, 259, 3152, 1984.
6. Owens, G. K. and Thompson, M. M., Developmental changes in isoactin expression in rat aortic smooth muscle cells *in vivo:* relationship between growth and cytodifferentiation, *J. Biol. Chem.*, in press, 1986.
7. Owens, G. K., Gordon, D., Loeb, A., and Thompson, M., Expression of smooth muscle specific isoactins in cultured vascular smooth muscle cells: relationship between growth and cytodifferentiation, *J. Cell Biol.*, 102, 343, 1986.
8. Larsen, D. M., Fujiwara, K., Alexander, R. W., and Gimbrone, M. A., Myosin in cultured vascular smooth muscle cells: immunofluorescence and immunochemical studies of alterations in antigenic expression, *J. Cell Biol.*, 99, 1582, 1984.
9. Gunther, S., Alexander, W., Atkinson, W. J., and Gimbrone, M. A., Functional angiotensin II receptors in cultured vascular smooth muscle cells, *J. Cell Biol.*, 92, 289, 1982.
10. Martin, W. and Gordon, J. L., Spontaneous and agonist induced Rb efflux from rabbit aortic smooth muscle cells in culture: a comparison with fresh tissue, *J. Cell. Physiol.*, 115, 53, 1983.
11. Smith, J. B. and Brock, T. A., Analysis of angiotensin-stimulated sodium transport in cultured smooth muscle cells from rat aorta, *J. Cell. Physiol.*, 114, 284, 1983.
12. Colucci, W. S., Brock, T. A., Gimbrone, M. A., and Alexander, R. W., Regulation of $alpha_1$-adrenergic receptor coupled calcium flux in cultured vascular smooth muscle cells, *Hypertension*, 6, I19, 1984.
13. Hartshorne, D. J. and Gorecka, A., Biochemistry of the contractile proteins of smooth muscle, in *Handbook of Physiology, Section 2, The Cardiovascular System*, Vol. 2, *Vascular Smooth Muscle*, Bohr, D. F., Somlyo, A. P., and Sparks, H. V., Jr., Eds., American Physiological Society, Bethesda, Md., 1980, 93.
14. Murphy, R. A., Mechanics of vascular smooth muscle, in *Handbook of Physiology, Section 2, The Cardiovascular System, Vol. 2, Vascular Smooth Muscle*, Bohr, D. F., Somlyo, A. P., and Sparks, H. V., Jr., Eds., American Physiological Society, Bethesda, Md., 1980, 325.
15. Somlyo, A. V., Bond, M., Berner, P. F., Ashton, F. T., Holtzer, H., and Somlyo, A. P., The contractile apparatus of smooth muscle: an update, in *Smooth Muscle Contraction*, Stephens, N. L., Ed., Marcel Dekker, New York, 1984, 1.
16. Murphy, R. A., Aksoy, M. O., Cohen, D. M., Fatigati, V., and Gerthoffer, W. T., Contractile apparatus and functional diversity in vertebrate smooth muscle, in *Vascular Neuroeffector Mechanisms: 4th Int. Symp.*, Bevan, J. A., Maxwell, R. A., Shibata, S., Fujiwara, M., Mohri, K., and Toda, N., Eds., Raven Press, New York, 1983, 37.
17. Dillon, P. F. and Murphy, R. A., High force development and crossbridge attachment in smooth muscle from swine carotid arteries, *Circ. Res.*, 50, 799, 1982.
18. Gabella, G., The force generated by a visceral smooth muscle, *J. Physiol. (London)*, 263, 199, 1976.

19. Barany, M., ATPase activity of myosin correlated with speed of muscle shortening, *J. Gen. Physiol.*, 50, 197, 1967.
20. Weisbrodt, N. W. and Murphy, R. A., Myosin phosphorylation and contraction of feline esophageal smooth muscle, *Am. J. Physiol.*, in press, 1985.
21. Kamm, K. E. and Stull, J. T., Myosin phosphorylation, force and V_0 in neurally stimulated tracheal smooth muscle, *Am. J. Physiol.*, in press, 1985.
22. Murphy, R. A., Gerthoffer, W. T., Trevethick, M. A., and Singer, H. A., Ca^{2+}-Dependent regulatory mechanisms in smooth muscle, in *Vasodilator Mechanisms*, Vanhoutte, P. M. and Vatner, S. F., Eds., S. Karger, Basel, 1984, 99.
23. Aksoy, M. O., Mras, S., Kamm, K. E., and Murphy, R. A., Ca^{2+}, cAMP, and changes in myosin phosphorylation during contraction of smooth muscle, *Am. J. Physiol.*, 245(14), C255, 1983.
24. Chatterjee, M. and Murphy, R. A., Calcium-dependent stress maintenance without myosin phosphorylation in skinned smooth muscle, *Science*, 221, 464, 1983.
25. Siegman, M. J., Butler, T. M., Mooers, S. U., and Davies, R. E., Calcium-dependent resistance to stretch and stress relaxation in resting smooth muscles, *Am. J. Physiol.*, 231, 1501, 1976.
26. Morgan, J. P. and Morgan, K. G., Calcium and cardiovascular function. Intracellular calcium levels during contraction and relaxation of mammalian cardiac and vascular smooth muscle as detected with aequorin, *Am. J. Med.*, 77, 33, 1984.
27. Gerthoffer, W. T. and Murphy, R. A., Ca^{2+}, myosin phosphorylation, and relaxation of arterial smooth muscle, *Am. J. Physiol.*, 245(14), C271, 1983.
28. Takano-Ohmuro, H., Obinata, T., Mikawa, T., and Masaki, T., Changes in myosin isoenzymes during development of chicken gizzard muscle, *J. Biochem. (Tokyo)*, 93, 903, 1983.
29. Burridge, K. and Bray, D., Purification and structural analysis of myosin from brain and other non-muscle tissues, *J. Mol. Biol.*, 99, 1, 1975.
30. Seidel, C. L., White, V., and Schildmeyer, L., Intervessel differences in smooth muscle myosin heavy chains, *Physiologist*, 26, A-12, 1983.
31. Rovner, A. S., Thompson, M. M., and Murphy, R. A., Detection of myosin heavy chain variants in smooth muscle by SDS polyacrylamide gel electrophoresis, *Fed. Proc.*, 43, 427, 1984.
32. Groschel-Stewart, U., Comparative studies of human smooth and striated muscle myosins, *Biochim. Biophys. Acta*, 229, 322, 1971.
33. Groschel-Stewart, U., Chamley, J. H., Campbell, G. R., and Burnstock, G., Changes in myosin distribution in dedifferentiating and redifferentiating smooth muscle cells in tissue culture, *Cell Tissue Res.*, 165, 13, 1975.
34. Larsen, D. M., Fujiwara, K., Alexander, R. W., and Gimbrone, M. A., Jr., Heterogeneity of myosin antigenic expression in vascular smooth muscle in vivo, *Lab. Invest.*, 50, 401, 1984.
35. Burridge, K., A comparison of fibroblast and smooth muscle myosins, *FEBS Lett.*, 45, 14, 1974.
36. Towbin, H., Staehelin, T., and Gordon, J., Electrophoretic transfer of proteins from polyacrylamide gels to nitrocellulose sheets: procedure and some applications, *Proc. Natl. Acad. Sci. U.S.A.*, 76, 4350, 1979.
37. Groschel-Stewart, U., Schreiber, J., Mahlmeister, C., and Weber, K., Production of specific antibodies to contractile proteins, and their use in immunofluorescence microscopy. I. Antibodies to smooth and striated chicken muscle myosins, *Histochemie*, 46, 229, 1976.
38. Vandekerckhove, J. and Weber, K., The complete amino acid sequence of actins from bovine aorta, bovine heart, bovine fast skeletal muscle, and rabbit slow skeletal muscle, *Differentiation*, 14, 123, 1979.
39. Fatigati, V. and Murphy, R. A., Actin and tropomyosin variants in smooth muscles. Dependence on tissue type, *J. Biol. Chem.*, 259, 14383, 1984.
40. Vandekerckhove, J. and Weber, K., At least six different actins are expressed in a higher mammal: an analysis based on the amino acid sequence of the amino-terminal tryptic peptide, *J. Mol. Biol.*, 126, 783, 1978.
41. Gabbiani, G., Schmid, E., Winter, S., Chaponnier, C., DeChastonay, C., Vandekerckhove, J., Weber, K., and Franke, W. W., Vascular smooth muscle cells differ from other smooth muscle cells: predominance of vimentin filaments and a specific α-type actin, *Proc. Natl. Acad. Sci. U.S.A.*, 78, 298, 1981.
42. Saborio, J. L., Segura, M., Flores, M., Garcia, R., and Palmer, E., Differential expression of gizzard actin genes during chick embryogenesis, *J. Biol. Chem.*, 254, 11119, 1979.
43. Devlin, R. B. and Emerson, C. P., Jr., Coordinate regulation of contractile protein synthesis during myoblast differentiation, *Cell*, 13, 599, 1978.
44. Bulinski, J. C., Kumar, S., Titani, K., and Hauschka, S. D., Peptide antibody specific for the amino terminus of skeletal muscle α-actin, *Proc. Natl. Acad. Sci. U.S.A.*, 50, 1506, 1983.
45. Pardo, J. V., Pittenger, M. F., and Craig, S. W., Subcellular sorting of isoactins: selective association of γ-actin with skeletal muscle mitochondria, *Cell*, 32, 1093, 1983.

46. Pollard, T. D., Cytoplasmic contractile proteins, *J. Cell Biol.*, 91, 1565, 1981.
47. McKenna, N., Meigs, J. B., and Wang, Y. L., Identical distribution of fluorescently labeled brain and muscle actins in living cardiac fibroblasts and myocytes, *J. Cell Biol.*, 100, 292, 1985.
48. Chamley-Campbell, J. H., Campbell, G. R., and Ross, R., Phenotype-dependent response of cultured aortic smooth muscle to serum mitogens, *J. Cell Biol.*, 89, 379, 1981.
49. Garrison, J. C. and Johnson, M. L., A simplified method for computer analysis of autoradiograms from two-dimensional gels, *J. Biol. Chem.*, 257, 13144, 1982.
50. O'Farrell, P. H., High resolution two-dimensional electrophoresis of proteins, *J. Biol. Chem.*, 250, 4007, 1975.
51. Gown, A. M., Vogel, A. M., Gordon, D., and Lu, P. L., A smooth muscle cell specific monoclonal antibody that recognizes smooth muscle actin isozymes, *J. Cell Biol.*, in press, 1985.
52. Libby, P. and O'Brien, K. V., Culture of quiescent arterial smooth muscle cells in a defined serum-free medium, *J. Cell. Physiol.*, 115, 217, 1983.
53. Ross, R., Nist, C., Kariya, B., Rivest, M. J., Raines, E., and Callis, J., Physiological quiescence in plasma-derived serum: influence of platelet-derived growth factor on cell growth in culture, *J. Cell. Physiol.*, 97, 497, 1978.
54. Baserga, R., *Multiplication and Division in Mammalian Cells*, Marcel Dekker, New York, 1976.
55. Chamley-Campbell, J. H. and Campbell, G. R., What controls smooth muscle phenotype? *Atherosclerosis*, 40, 347, 1981.
56. Rovner, A. S., Murphy, R. A., and Owens, G. K., Expression of smooth muscle and nonmuscle myosin heavy chains in cultured vascular smooth muscle cells, *J. Biol. Chem.* 1986, in press.
57. Anderson, J. M., Gimbrone, M. A., Jr., and Alexander, R. W., Angiotensin II stimulates phosphorylation of the myosin light chain in cultured vascular smooth muscle cells, *J. Biol. Chem.*, 256, 4693, 1981.
58. Yau-Young, A. O., Shio, H., and Fowler, S., Growth, biochemistry, and morphology of isolated rabbit aortic smooth muscle cells maintained in the presence or absence of serum, *J. Cell. Physiol.*, 108, 461, 1981.
59. Schubert, D., Harris, A. J., Devine, C. E., and Heinemann, S., Characterization of a unique muscle cell line, *J. Cell Biol.*, 61, 398, 1974.
60. Olson, E. N., Caldwell, K. L., Gordon, J. I., and Glaser, L., Regulation of creatine phosphokinase expression during differentiation of BC_3H_1 cells, *J. Biol. Chem.*, 258, 2644, 1983.
61. Karnovsky, M. J., Endothelial-vascular smooth muscle cell interactions, *Am. J. Pathol.*, 105, 200, 1981.

Chapter 6

GROWTH FACTORS FOR VASCULAR SMOOTH MUSCLE CELLS

Daniel F. Bowen-Pope, Mark W. Majesky, and Russell Ross

TABLE OF CONTENTS

I. INTRODUCTION

The size, elasticity, and strength of the large arteries is largely determined by the smooth muscle cells (SMC) of the tunica media and by the macromolecules (especially collagen and elastin) synthesized and deposited by these cells. Weakness of, or damage to, this layer can result in the formation of aneurysms or hemorrhages. Conversely, excessive growth of SMC in the tunica intima can impede the flow of blood, as in atherosclerosis[1,2] and hypertension.[3] How is the architecture and mass of the vessel wall regulated? This chapter will discuss some of the information which has been gained from studying vascular SMC in culture. The information should be evaluated with the following considerations in mind.

A. The In Vivo vs. Cell Culture Environments

The environment of an SMC in culture may be very different from that of an SMC in vivo. Some of these differences may affect the way an SMC would respond to a given stimulus. The culture environment itself can vary tremendously between different experimental protocols. Certain protocols may predispose the cells to respond to specific variables. When stimulation of cell proliferation is studied, a low basal (control) rate of proliferation must first be established. This has been variously accomplished by allowing the cultures to proliferate until growth is slowed by some combination of cell density and growth factor limitations, by reducing the concentrations of serum added to the nutrient medium, or by replacing the whole blood serum with plasma-derived serum or by a defined mixture of supplementary nutrients and factors. Each different protocol could predispose the test cells to respond to a different "growth factor". For example, single amino acids can become growth stimulatory when assayed using SMC cultured in medium depleted specifically of that amino acid. Studies using cultured cells can thus indicate which factors *could* stimulate SMC proliferation, but do not in themselves indicate that changes in levels of these factors occur, or are important, in vivo. Indeed, none of the factors studied using cultured SMC have yet been unambiguously demonstrated to function as growth factors for SMC in vivo. As an aid to evaluating the significance of results obtained with different putative growth factors in culture, we will briefly describe the specific culture system used to obtain the result and will mention the arguments adduced for a physiological role for the factor.

B. Phenotypic Modulation and Heterogeneity

The properties of SMC in culture may differ considerably from those displayed by their in-vivo progenitors. Such differences could arise through phenotypic "modulation" and/or through selection of subpopulations. SMC in the arteries of normal adults (and during the first days in culture) are characteristically well filled with thick and thin myofilaments and dense bodies and contain a poorly developed Golgi apparatus and rough endoplasmic reticulum.[4-6] When SMC are isolated from the tunica media by enzymatic digestion and put into culture, they lose most of their observable myofilaments and develop an extensive Golgi and rough endoplasmic reticulum.[4,7] This process will be discussed extensively in other chapters of this book (see Chapters 4 and 13). Observations in several systems suggest that these morphological changes are accompanied by changes in responsiveness to mitogens. Jarmolych et al.[5] and Fritz et al.[6] observed that during the first 2 days after explanting, very few (0 and 0.4%) of the SMC in explants of pig and rabbit thoracic aortas incorporated ^{3}H-thymidine during a 2-hr labeling period. During the next 2 days, increasing numbers of labeled cells (11% on day 4) were observed. During the same period, many of the cells were undergoing the morphological changes described above, and electron microscope autoradiography indicated that *only* modulated (synthetic phenotype) cells incorporated ^{3}H-thymidine.[6] A comparable phenomenon has been observed using enzymatically isolated SMC. The percentage of pig and monkey SMC incorporating ^{3}H-thymidine increases rapidly at about 7 days after plating.[4] This correlates with the appearance of cells with a well-developed synthetic phenotype. Enzymatically isolated rat SMC undergo a similar modulation at 2 to 4 days after isolation.[7] Only cells in the synthetic phenotype were observed by electron microscopic autoradiography to be incorporating ^{3}H-thymidine.

Chamley-Campbell et al.[4] reported conditions under which enzymatically isolated SMC could be kept in the contractile phenotype. However, the culture procedures routinely used by virtually all investigators yield cells in the synthetic phenotype, and the observations reported in this chapter may be applicable only to SMC with this phenotype. If the "synthetic phenotype" was expressed only in the tissue culture environment, this could compromise the utility of studying SMC in culture. There is reason to speculate, however, that transitions between contractile and synthetic phenotypes occur in vivo as well. SMC in the aortas of immature rats[8-10] and after experimental injury[11] have been described as having the morphological characteristics of the synthetic phenotype. Explants cultured from rabbit[12] and rat[13] aortas, which were injured by balloon catheter deendothelialization 4 to 7 days before sacrifice, showed more rapid outgrowth of cells, possibly as a result of prior modulation to the synthetic phenotype. It seems possible that proliferation of SMC in vivo, as well as in culture, is controlled at two levels: (1) by a transition into (or out of) a mitogenically competent synthetic phenotype and (2) by the presence (or absence) of "growth factors" active on the synthetic phenotype cells. Each level of control may be exerted by a different set of factors. Thus, the growth factors in platelets, which are so active on synthetic phenotype SMC, do not seem to influence the transition of pig and monkey SMC from contractile to synthetic phenotype.[14] Thyberg et al.,[7] however, reported that modulation of rat SMC is more rapid in whole blood serum (WBS) than in plasma-derived serum (PDS) and that treatment of WBS with antibodies against the platelet-derived growth factor (PDGF) delays modulation.

A second consideration which complicates analysis of SMC responses both in vivo and in culture is the possibility that the cells in the tunica media are phenotypically heterogeneous. The tunica media of the large arteries of many species is usually considered to consist entirely of a single cell type so that cells obtained from this source are all considered to be SMC.[15] However, in some species the tunica media contains more than one morphologically[16] or histochemically[17] distinguishable cell type. If subsets of SMC with different growth properties existed within the artery wall, this would cause problems for cell culture analysis of growth in cell culture. On one hand, the cells studied in culture could represent a subset of the SMC in the vessel wall that have been selected for their ability to proliferate well in culture. The information obtained in culture would then be applicable to only a subset of the SMC in the vessel wall. On the other hand, if one subtype is *not* preferentially obtained in culture, then analysis of response of cultured cells to growth factors will be complicated by the possibility that the different growth factors are stimulating different subpopulations. This particularly complicates analysis of the effects of simultaneous addition of more than one factor, since each factor could be acting on either the same or different subtypes. The cloning of SMC offers a possible approach toward investigating the possibility, and consequences, of cell heterogeneity in the vessel wall.

C. Use of the Term "Growth"

The final caveat is that the term "growth" in "growth factor" will be considered here to include all processes which would contribute toward an increase in the mass of cells in the tunica media. As measured in cell culture, these processes include an increase in the number of cells synthesizing DNA (determined by ^{3}H-thymidine incorporation and flow microfluorometry), an increase in the number of cells, an increase in the amount of protein or DNA per culture, or an increase in colony size. These processes are not necessarily equivalent mechanistically; e.g., an increased rate of entry of cells into the DNA synthetic phase of the cell cycle would not lead to an increase in cell number if karyokinesis did not occur (Owens et al.[3] have shown that vascular SMC can become polyploid) or if cell survival declined. Conversely, a change in cell survival could lead to a change in cell number even if the rate of entry into S was not affected.

II. GROWTH FACTORS IN BLOOD

A. Analysis of Blood Constituents

Whole blood serum (WBS usually ambiguously referred to simply as "serum") is the archetypal "growth factor" for most cultured mesenchymal cells, including vascular SMC. It has been used to supplement nutrient medium since the earliest days of cell culture. It was well recognized that the serum could be supplying "growth factors", as well as the essential nutrients and vitamins that were not included in the chemically defined portion of culture medium. The problem was to determine the identity of these factors. Two approaches have been especially fruitful:

1. Serum has been obtained from animals in different nutritional, hormonal, or metabolic states and tested for differential ability to stimulate the test cells. If a difference is found, knowledge about the state of the animals is then used to suggest a candidate for the key variable from among the spectrum of factors known to be present in serum. Ideally, this choice is confirmed by verifying that specific neutralization of the factor eliminates the difference and that specific addition of the factor reproduces the difference. This approach is obviously particularly well suited to evaluate the role of already well-characterized growth factors (e.g., insulin — see later).
2. A second approach has been to start with serum from normal animals and attempt to isolate the active components by biochemical techniques, using growth-promoting activity as an assay to identify the active fractions. This approach is necessary for characterizing growth factors that have not been previously identified or purified and has resulted in the discovery of several new factors (e.g., Somatomedin, PDGF). One specific form of analysis of blood into components has both biochemical utility and physiological significance — the division into plasma and cell-derived constituents.

B. Growth Factors in Platelets

1. Plasma-Derived Serum vs. Whole Blood Serum

Although blood derivatives have long been used to supplement nutrient medium, whole blood is very rarely, if ever, used. The relationships between some of the commonly used blood derivatives are diagramed in Figure 1. Carrell and Burrows[18] placed tissue explants in plasma, which rapidly coagulated around the tissue and provided a semisolid environment into which cells migrated and proliferated. Current cell culture techniques are hindered by the presence of the clot, so the coagulation cascade is activated in vitro prior to exposure of the plasma to cells, and only the supernatant "serum" is used. If the cellular elements of blood are removed (usually by centrifugation) before coagulation, the fluid is designated "plasma-derived serum" (PDS). If not, it is properly designated "whole blood serum" (WBS) but is usually referred to

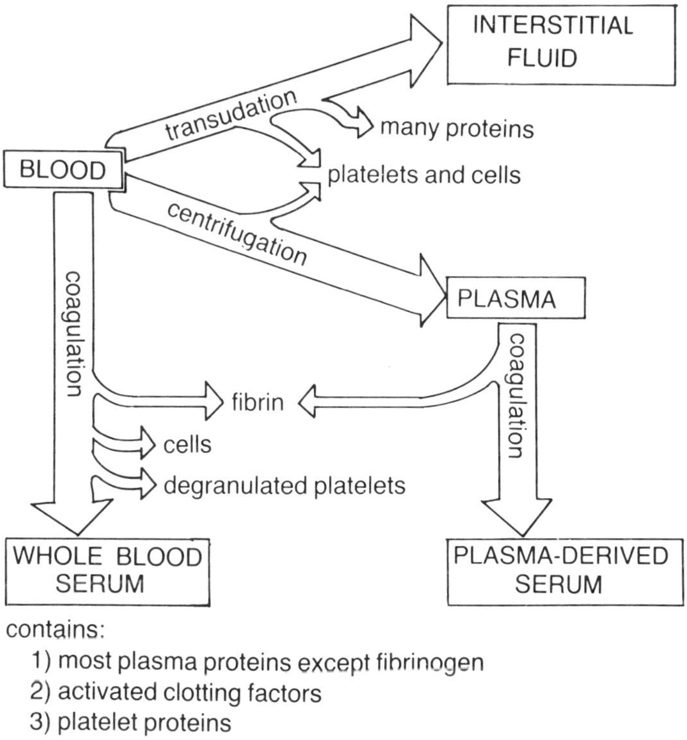

FIGURE 1. Derivation of different fluids from blood. (From Bowen-Pope, D. F., in *The Cell Surface in Development and Cancer,* Browder, L., Ed., Plenum Press, New York, in press. With permission.)

simply as "serum". As will be described below for SMC, WBS was found to support more rapid cell proliferation than did PDS.[19-21] This difference was found to be due to the release of platelet components into serum as a consequence of the activation of platelet degranulation during coagulation.[20-23] Several growth factors are now known to be released from platelets: PDGF, type-beta transforming growth factor (β-TGF), and a factor related to epidermal growth factor (EGF). The effects of individual factors on SMC growth will be described later. Although such studies provide more information as to which platelet factors might be capable of stimulating SMC, and to what extent, the use of WBS is still informative. Since WBS contains plasma and all of the platelet growth factors in the proportions in which they are released from platelets, it may provide the closest available approximation to the mixture of the growth factors to which SMC are actually exposed in vivo at sites of vascular injury.

The greater potency of WBS than of PDS in stimulating SMC growth has been demonstrated for SMC from many species using both homologous and heterologous sera. Ross et al.,[21] Rutherford and Ross,[24] and Burke and Ross[25] showed that monkey (*Macaca nemestrina*) thoracic SMC responded to 5% monkey WBS with greater rates of proliferation and percent labeled nuclei, and that the potency of PDS could be increased to that of WBS by adding material released by platelets in response to thrombin.[21] Similar reports of differential response to homologous PDS and WBS have been made for human arterial SMC[21,26,27] and rabbit thoracic SMC.[28] PDS and WBS retain their differential stimulatory capacity when tested on heterologous SMC, e.g., human sera tested on rat aortic SMC[7,29] or monkey aortic SMC.[30]

Although this chapter focuses on growth factors, the relative lack of potency of PDS as a growth factor is equally important for understanding growth control. SMC in the artery wall of normal adult animals are healthy but not rapidly proliferating.[31] These cells are exposed to the subset of blood components ("interstitial fluid") which pass through the vascular endothelium. As is diagramed in Figure 1, interstitial fluid differs from both PDS and WBS in that it does not contain activated clotting factors. It shares with PDS the absence of growth factors from platelets. Given the relative ease with which PDS can be prepared (e.g., see Ross et al.[21]), it has been proposed as a useful substitute for interstitial fluid in bringing SMC into a state of healthy quiescence in culture.[30] In some cases, PDS has been further processed by incubation with carboxymethyl sephadex to remove cationic factors, including any PDGF which might have been released from the platelets before their removal.[30,32] Now that specific assays for PDGF are available, it has been found[33-35] that carefully prepared PDS does not contain detectable PDGF. Since treatment with carboxymethyl sephadex removes many cationic proteins from serum and the accompanying dialysis and ultrafiltration steps can cause loss of soluble nutrients, this procedure is no longer routinely performed. As an alternative to the use of PDS, Libby and O'Brien[36] have developed a chemically defined culture medium which maintains bovine arterial SMC in a stable quiescent state (stable DNA content per culture). This medium is based on a 1:1 mixture of Dulbecco's modified Eagle's medium and Ham's medium F-12 supplemented with 10^{-6} M insulin, 5 μg/mℓ transferrin, and 0.2 mM ascorbate.

Use of chemically defined and complex (e.g., PDS) supplements can each contribute toward an analysis of SMC response to different growth factors. Use of a simple defined medium might not permit cells to respond to certain factors since required "cofactors" which may normally be present in blood or in interstitial fluid are not present. A striking example of this which has been observed using fibroblastoid cells, but not yet using SMC, is the synergism between PDGF, EGF, and β-TGF in stimulating the proliferation of cells suspended in soft agar (see later). Alone, none of these factors stimulates efficient growth in suspension. In combination, they are very potent.[37,38] The original observations of soft agar growth-promoting activity were possible because the assay was performed in nutrient medium supplemented with subthreshold concentrations of WBS that contributed the necessary cofactors.[39,40] Now that the synergism between these three factors has been recognized, the mechanism(s) of their interactions can be investigated using defined medium without serum. In general, it seems that chemically defined medium may be very useful for studying mechanisms, but that it is susceptible to generating both false negatives (due to absence of a "cofactor" present in vivo) and false positives (due to presence of a factor not present in blood or present at much lower concentrations) when used to investigate which factors may actually function in vivo.

2. Platelet-Derived Growth Factor

The original observations of the differences between PDS and WBS in stimulating proliferation of cultured chicken embryo cells,[19] monkey SMC,[21] mouse "fibroblastoid" 3T3 cells,[20] and a glial cell line[22] and the observation that the difference could be accounted for by growth factors released from platelets[20-22] led eventually to the purification from human platelets[41,42] and from human platelet-rich plasma[43,44] of a mitogenic 30,000-dalton, polypeptide factor designated PDGF. The fact that four laboratories, using different growth assays and biochemical procedures, independently purified a single growth factor from human blood by-products is consistent with the proposal that PDGF is one of the major growth-promoting factors in human blood. The amount of PDGF in human WBS has now been determined by specific radioreceptor assay (15 ng/mℓ,[33] 17.5 ng/mℓ[34]) and by radioimmunoassay (50 ng/mℓ,[45] 50 ng/mℓ,[34] 15.5 ng/mℓ[46]). Since the ED_{50} for stimulation of thymidine incorporation by several connective tissue cell types, including SMC, is about 1 to 3 × 10^{-11} M (0.3 to 1 ng/mℓ[47] and see later), the PDGF content of 5 to 10% human WBS would make a very substantial contribution to the growth-promoting activity of WBS. The preparation of monospecific antisera against human PDGF has allowed the relative contribution of PDGF to the total mitogenicity of human WBS to be determined by specific neutralization. Our own (unpublished) studies and those of Heldin et al.[48] indicate that PDGF accounts for approximately 50% of the stimulatory activity of human WBS for 3T3 and glial cells, respectively. Nilsson et al.[49] confirmed that antiserum against PDGF removed a majority of the ability of human WBS to stimulate proliferation of aortic SMC from 5-day-old rats, but reported that this was not true for adult SMC.

PDGF-like molecules have been found by radioreceptor assay in WBS obtained from all chordates tested,[33] but the concentration of PDGF present can vary considerably from species to species. In our own studies, we have found that mouse WBS also contains high levels of PDGF (≃10 ng/mℓ) but that baboon WBS contains only 2.7 ng/mℓ,[35] and commercial calf serum contains even less (≃0.75 ng/mℓ, our unpublished data). Since the growth-promoting capacity of human and calf WBS are comparable, it would seem that PDGF contributes less to the growth-promoting activity of calf serum than it does to that of human WBS. The relative paucity of PDGF in calf serum could help to explain why PDGF was not obtained in the early attempts to fractionate serum (bovine) into its mitogenically active components (reviewed by Temin et al.[50]).

Purified PDGF has been shown to stimulate the growth of many mesenchymal cell types, including vascular SMC from several species. Incorporation of ^{3}H-thymidine by monkey (*Macaca nemestrina*) and human arterial SMC is stimulated with an ED_{50} of approximately 10^{-11} *M*.[47] Mitogenic response was paralleled by specific high-affinity binding of ^{125}I-PDGF.[47] Parallel stimulation of ^{3}H-thymidine incorporation and specific binding of ^{125}I-PDGF were also reported for bovine aortic SMC,[51] in that case with a somewhat higher ED_{50}. PDGF stimulates proliferation and increases ^{3}H-thymidine labeling indices of adult rat aortic SMC in a base of 0.5 to 1% fetal bovine serum[52] and in serum-free medium.[49,53] The latter authors speculate that previous reports of the need for plasma components for stimulation of certain cell types by PDGF may have resulted from the use of nutritionally inadequate basal medium (i.e., DMEM rather than MCDB 104). Alternatively, the cultured rat SMC could themselves be synthesizing the key factors normally provided by plasma (see Atkinson et al.[54] and Clemmons[55] for evidence that cultured fibroblasts synthesize Somatomedin C). The stimulation of rat SMC by PDGF was inhibited by prostaglandin PGE_1 at 10 ng/mℓ and by PGF_2 and PGD_2 at higher concentrations,[53] confirming previous observations that these prostaglandins inhibited ^{3}H-thymidine incorporation by rabbit aortic SMC cultured in 10% fetal bovine serum[56] and reduced the cloning efficiency of guinea pig aortic SMC in 10% fetal bovine serum.[57] Thyberg et al.[7] found that adult rat aortic SMCs respond to the anionic growth factors in human platelets as well as to PDGF so that the effects of WBS on SMC growth should not all be attributed to PDGF.

The responsiveness of cells to PDGF seems to be mediated through expression of cell-surface receptors for PDGF[47,51,58,59] (reviewed in Bowen-Pope et al.[60]). PDGF receptors have been measured on SMC from human aorta,[47] monkey aorta,[47,61] porcine aorta,[58] rat aorta,[62] and bovine aorta.[51,60,63] PDS contains a high-molecular-weight proteins(s) which prevents binding of PDGF to this cell-surface receptor.[34,36,64,65] We calculate that if a small amount of PDGF was released into plasma, about 75% of this would be present as an inactive complex with the binding protein. The biological effects of complex formation in culture or in vivo are still speculative. In culture, the combination of very high affinity receptors and very low working concentrations of PDGF contributes toward a situation in which the added PDGF can be depleted during the course of a mitogenesis experiment. By preventing the cultures from squandering their PDGF during the first few hours, the binding protein could increase the mitogenic efficacy of PDGF. In vivo, a similar situation could exist or, alternatively, the binding protein could function entirely as a negative mediator by inactivating PDGF, which doesn't very quickly bind to PDGF receptors and/or by affecting the way PDGF is cleared from the plasma compartment by degradation systems.[35]

3. Epidermal Growth Factor

Despite its name, EGF is a potent mitogen for many connective tissue cell types, including SMC (reviewed in Adamson and Rees[66]). EGF was first discovered because of its presence in huge quantities within the mouse salivary glands.[67] Although EGF is now one of the best characterized growth factors and has served as a model for many of the biochemical and cell biological studies performed with PDGF, its source and physiological roles in humans are far from clear. It is not, for example, present in high concentrations in the human salivary gland. Even in the mouse, the relationship between the high concentrations in salivary glands and the concentrations seen in other tissues is far from clear, since surgical removal of the salivary gland does not alter levels of EGF in other tissues or in blood.[68] While investigating interactions between the receptors for PDGF and for EGF, we made measurements of EGF levels in blood, distinguishing what was in platelets from what was in plasma. We found that all EGF detectable in blood appeared to be present within the platelets.[69] Oka and Orth[70] made a similar observation and concluded that platelets are the major repository of EGF in man. It is likely that EGF-like molecules constitute some of the anionic growth factor activity found in WBS.

EGF can stimulate growth of SMC under certain conditions. At 100 ng/mℓ, EGF increases the apparent saturation density of adult bovine aortic SMC in a base of either 2.5% PDS or 2.5% WBS.[71] In serum-free medium supplemented with 250 μg/mℓ of high-density lipoprotein (HDL), EGF stimulates proliferation with an ED_{50} of 5 ng/mℓ and an optimum at 50 ng/mℓ.[72] ^{3}H-thymidine incorporation by confluent cultures of calf aortic SMC is increased tenfold by the addition of 10 ng/mℓ EGF to serum-free medium.[73] Rat abdominal aortic SMCs in a base of 0.5 to 1% FBS were reported to show increased proliferation in response to 10 ng/mℓ EGF only if partially purified Somatomedin A (10 μg/mℓ) was also added.[52]

4. Class β-Transforming Growth Factor

Transforming growth factors (TGF) are operationally defined as substances which promote the growth in suspension of cultured cells that otherwise require attachment to a substratum.[74] Ability to grow in soft agar is one of the characteristics of oncogenically transformed cells which most reliably distinguishes them from their nontransformed counterparts.[75] Class α-TGF were first found to be secreted by cultured transformed cells. These TGF resemble EGF in that they bind to the EGF receptor[76] and are related to EGF in amino acid sequence.[77] Class β-TGF do not bind to the EGF receptor and depend on the presence of EGF in order to display optimal ability to promote growth in soft agar.[37] Surprisingly, the normal tissue richest in class β-TGF seems to be the blood platelets.[39,40] Assoian et al.[38] have provided evidence that the β-TGF in platelets may be strongly supported in its activity by EGF and PDGF together, so that a platelet contains a complete set of three growth factors needed to support optimal growth in soft agar: PDGF, EGF, β-TGF.

To date, β-TGF has been studied almost exclusively using cultured cell lines. Nevertheless, β-TGF receptors seem to be expressed by almost every cell type and may play a role in regulating SMC growth in vivo. The environment of a smooth muscle cell in a damaged artery, and possibly in a normal artery as well, is not identical to that of an attached cell in monolayer culture and may share some characteristics with suspension in soft agar. A "transforming growth factor" would then be necessary to promote optimal cell proliferation. The name "transforming growth factor" should probably be viewed as reflecting the historical circumstances of its initial discovery rather than as implying a unique involvement in oncogenic transformation. This possibility is supported by the fact that β-TGF is encoded by a cellular gene. It thus seems that oncogenic transformation can activate cellular genes encoding for growth factors related to three of the growth factors found in platelets: PDGF (e.g., Bowen-Pope et al.[78]), EGF (e.g., Todaro et al.[76]), and β-TGF,[79] thereby, presumably, making the tumor cells independent of exogenous mitogens.

5. Platelet Factors In Vivo

Since most of the experimental work with the above-mentioned platelet factors has been done in culture, the exact role(s) of these factors in vivo has not been established. Nevertheless, several pieces of evidence favor the hypothesis that factors released from platelets stimulate the proliferation of connective tissue cells at sites of tissue damage:

1. Growth factors from platelets have been localized to the α-granules by studying platelets from patients with α-granule-specific defects,[80] by fractionating subcellular organelles[23] and by studying the differential release of granule components by different platelet-activating agents.[26] Release of granule proteins from the platelet is exquisitely sensitive to environmental stimuli, including contact with exposed subendothelial connective tissue.[81,82] This provides a mechanism for the topical application of growth factors to sites at which a proliferative response would be useful. Delivery of platelet factors in this fashion has been confirmed by Goldberg et al.,[83] who reported that large amounts of PF4, another α-granule protein, have penetrated into the vessel wall within 10 min after balloon catheter injury.
2. Inhibition of platelet function reduces the proliferative response of SMC to experimental injury, presumably by preventing the local release of growth factors.[84-87]

C. Growth Factors in Plasma

1. Insulin and Insulin-like Growth Factors

Insulin and insulin-like growth factors (IGF) constitute a family of evolutionarily related polypeptide factors which can stimulate both growth and insulin-like metabolic effects in several mesenchymal cell types.[88,89] It seems likely that the mitogenic activity of insulin is mediated via the IGF receptors, which can be occupied by insulin, albeit with a much lower affinity than by IGF. This may explain why the mitogenic effects of insulin on cultured cells are often seen only when very high concentrations of insulin are used.

Bovine SMC are not stimulated by 5 μg/mℓ insulin in a base of medium 199 + 0.1% human serum albumin.[90] In a 1:1 mixture of media DV plus F12, 10^{-6} *M* insulin does diminish loss of cell protein and DNA from cultures of bovine SMC previously grown in serum-containing medium. When 5 μg/mℓ transferrin and 0.2 m*M* ascorbate were included, the cultures showed a positive protein balance and did not lose DNA.[36] Gospodarowicz et al.[72] reported that proliferation of bovine SMC in serum-free DME medium supplemented with HDL was stimulated by insulin with an apparent ED_{50} of 150 ng/mℓ and maximally at 2.5 ng/mℓ and was also stimulated by 10 ng/mℓ of Somatomedin-C (IGF-I).

In serum-free medium, insulin at 1 mU/mℓ stimulated a small increase in ^{3}H-thymidine incorporation by rat aortic SMC.[91] In a base of 0.5 to 1% fetal bovine serum, insulin stimulated proliferation and ^{3}H-thymidine incorporation by rat aortic SMC.[52,92] Maximal stimulation was achieved at 500 μU/mℓ, but was only about 20% as great as was achieved with 10% calf serum.[92] Insulin at 100 ng/mℓ was less effective than other members of the "insulin family", including partially purified Somatomedins A and C (10 μg/mℓ), multiplication stimulating activity (0.5 μg/mℓ), and IGF-I (50 ng/mℓ).[52] Proliferation of human arterial SMC in a base of 6% fetal calf serum is also stimulated by insulin, with significant stimulation seen at 10 μU/mℓ and a maximum effect seen at 10,000 μU/mℓ.[93] The maximum effect of insulin was again considerably less than that of 10% fetal bovine serum.

Koschinsky et al.[94] reported that serum from diabetic patients promoted better proliferation of human SMC than did serum from normal or treated diabetics. This confirms the observation by Ledet[95] that diabetic serum promotes more rapid growth of colonies from explants of rabbit aorta. However, increased levels of insulin in diabetic serum did not seem to account for the increased efficacy of the diabetic serum since addition of up to 2000 μU/mℓ of insulin to the normal serum did not stimulate growth.[95]

2. Growth Hormone

Ledet[95] has reported that diabetic serum stimulates more rapid growth of colonies of rabbit aortic SMC than does normal serum. Normal serum can be made more potent by the addition of growth hormone,[95,96] but not by the addition of insulin[95] or glucose.[96] Antiserum against growth hormone reduces the potency of diabetic serum to that of normal serum. The amount of growth hormone required to reproduce the extra stimulation by diabetic serum is in the same concentration range (0.2 ng/mℓ) as could be found in diabetic serum since growth hormone levels in diabetics are elevated three- to fourfold, from 0.4 to 1.4 ng/mℓ.[97] In vivo, considerably less aortic SMC proliferation occurs in response to balloon catheter de-endothelialization in rats that have been hypophysectomized than in control rats.[98,99] This does not seem to be due to deprivation of growth hormone, however, since replacement therapy with growth hormone alone (or together with glucocorticoids, mineralocorticoids, and thyroxine) did not restore the SMC proliferative response.[99]

3. Lipids and Lipoproteins

Several key observations demonstrating that lipoproteins can stimulate growth under certain circumstances have been made using the following culture system.[100-102] Explants of rabbit or Rhesus monkey thoracic aorta are cultured for 5 to 6 months in medium containing 10% normal serum. During this period, SMC migrate out from the explants and from colonies. Test sera are then substituted for 50% of the normal serum or test lipoproteins are added, and the growth response is measured as an increase in colony size. Stimulation was obtained with low-density lipoprotein (LDL) from hyperlipemic serum, but not with LDL from normolipemic serum or with HDL from either serum.[100] The differences remain even when the LDL from the two sera are used at a comparable cholesterol content.[101] Fless et al.[102] used this system to analyze LDL subfractionated by density gradient ultracentrifugation. The "larger LDL species" (molecular weight greater than 3.0×10^{-6}, density greater than 1.030 g/mℓ) from both normal or hyperlipemic monkeys or from some normal human donors was mitogenic. The "smaller LDL" (molecular weight = 2.6 to 2.7×10^{-6}) from normal monkeys or from some normal human donors was not mitogenic. "LDL-III", the Rhesus monkey equivalent of human Lp(a), was also mitogenic, but only after removal of most of its sialic acid moieties.

Using a culture system in which initial rates of proliferation of porcine aortic SMC were measured in a base of 1.5% porcine serum, Brown et al.[103] determined the dose response to several species of lipoproteins: HDL_c, HDL_2, LDL, and VLDL, listed in order of increasing maximum potency. Maximum effects of each were achieved at 10 mg/dℓ of lipoprotein cholesterol, with all responses being less than that produced by 10% porcine serum. The source of the lipoproteins (normal vs. hyperlipemic serum) did not make a difference. At triglyceride levels above 15 mg/dℓ, VLDL was reported to lead to foam cell formation. Whole hyperlipemic serum has been reported to be more stimulatory than normolipemic serum for rat,[91] rabbit,[104] and pig and monkey[14] SMC. Extraction of lipids from normolipemic serum with acetone/ethanol/ether reduced its ability to stimulate ^{3}H-thymidine incorporation by rat aortic SMC.[91] However, the extraction procedure could be affecting serum components in addition to lipoproteins, and certain fatty acids are essential nutrients for cultured mammalian cells. Specific removal of LDL and most HDL from normolipemic monkey WBS permitted proliferation of monkey SMC, though to a lesser extent than did reconstituted serum.[105]

In some culture systems, LDL seems to be toxic for SMC. Hessler et al.[106] reported that LDL from normolipemic human serum caused a decrease in the number of human SMC per culture, as well as in ^{3}H-thymidine labeling index, over a period of 3 days. By contrast, HDL was never toxic and prevented the toxicity of LDL. Gospodarowicz et al.[72] confirmed that LDL was toxic for cultured bovine aortic SMC and found that HDL was stimulatory at 250 μg/mℓ and used it to supplement DME medium in preparing a serum-free culture medium.

4. Other Factors in Plasma

Among the other factors present in plasma which have been tested for ability to stimulate growth of cultured SMC are the following: Thrombin (1 μg/mℓ), hydrocortisone (5×10^{-5} *M*) and transferrin (10 μg/mℓ) were reported to stimulate proliferation of rat abdominal aortic SMC in a base of 0.5 to 1% fetal bovine serum.[52] Ascorbate (50 μg/mℓ) in a base of 10% fetal bovine serum increased growth rate and plating efficiency, but did not alter the saturation density of bovine aortic SMC.[107] Libby and O'Brien[36] found that ascorbate (0.2 m*M*) in combination with insulin and transferrin could prevent the cell protein and DNA loss, which occurs when cultures of calf aortic SMC are transferred from medium supplement with serum to serum-free medium. The steroid hormones ethinylesteradol, estrone, and estriol at 10^{-8} *M* did not affect ^{3}H-thymidine incorporation by rat aortic SMC,[91] nor did thyroxine or several steroids or vasoactive hormones affect the survival of bovine aortic SMC in serum-free medium.[36]

III. GROWTH FACTORS FROM CULTURED CELLS

A. Growth Factors from Endothelial Cells

Cultured endothelial cells from several species (human, pig, bovine) have been shown to produce a growth factor(s), endothelial cell-derived growth factor (ECDGF), which can stimulate several connective tissue cell types.[108-110] ECDGF produced by bovine aortic endothelial cells cultured in serum-free medium stimulated a large increase in the ^{3}H-thymidine labeling index of bovine aortic SMC in serum-free medium.[108] ECDGF collected in 5% PDS supported proliferation of bovine aortic SMC as well as did 5% WBS.[108] ECDGF produced by bovine aortic endothelial cells cultured in serum-free medium or in 1% PDS also stimulated proliferation of human aortic SMC.[109] Serum-free medium conditioned for 1 day was found to be about two thirds as effective as 5% bovine serum.[109]

ECDGF is probably a mixture of several different factors. One of these factors can bind to the PDGF receptor and is recognized by antiserum prepared against pure

PDGF.[63] This PDGF-like factor accounts for approximately 25% of the mitogenic capacity of human umbilical vein ECDGF when tested by ability to stimulate ^{3}H-thymidine incorporation by 3T3 cells.[63,78] DiCorleto[111] has recently separated the PDGF-like and PDGF-unlike factors in bovine aortic ECDGF by cation exchange chromatography and confirmed that at least two biochemically distinct growth factors are present. Growth factor production by vascular endothelial cells thus resembles growth factor release by platelets in that both sources yield multiple factors. Since PDGF is a potent mitogen for SMC (see earlier), it is likely that the PDGF-like component of ECDGF will be active on SMC. The relative contribution of the PDGF-like component to the overall potency of ECDGF will depend on the susceptibility of the test cells to the other components of ECDGF and will not necessarily be comparable to the 25% value obtained, using 3T3 cells as the test cells. Endothelial cells from different vessels or from different species may also make different proportions of the PDGF-like and other components and may alter the proportion of PDGF-like components under different circumstances. Although we are not discussing growth inhibitors in this chapter, it should be noted that cultured vascular endothelial cells have also been reported to produce substances which inhibit the proliferation of SMC[110] (see also Chapters 4 and 7). Castellot et al.[110] have presented evidence that this inhibitory factor is a heparin-like molecule. Heparin from other sources has been demonstrated to inhibit the proliferation of SMC that occurs in response to vascular injury.[112]

It is not known whether endothelial cells in vivo release PDGFc. However, it has been reported that 3.7-kb mRNA hybridizing with a *sis* cDNA probe can be detected in endothelial cells scraped from human umbilical veins and from bovine aortas.[113] The *sis* gene encodes a protein homologous to one of the two related polypeptides found in preparations of pure PDGF and is probably the cellular gene for this PDGF peptide. The presence of *sis*-hybridizing mRNA in endothelial cells scraped directly from tissues is consistent with some expression of this PDGF gene in vivo. The bovine aortic and human umbilical vein endothelial cells scraped from vessels yielded approximately 1.3 and 10%, respectively, as much *sis*-hybridizing RNA as did the same cell types in culture, suggesting that endothelial cells in undamaged vessels in vivo produce less PDGFc than do their cultured counterparts. PDGFc production by cultured endothelial cells can be stimulated further by several agents. Fox and DiCorleto[114] have reported that tumor promoters and endotoxin can increase the rate of secretion by about twofold. Physiological concentrations of thrombin (1 to 10 units/mℓ) can increase rates of PDGFc secretion by HUVE by at least threefold.[115] The thrombin stimulation is not associated with decreases in the intracellular levels of PDGFc, suggesting that thrombin is not stimulating release of preformed, active PDGF. The effect of thrombin may thus be on the rate of synthesis of PDGFc by endothelial cells or on the rate of processing of an inactive PDGF processor to a secreted active PDGFc. These results suggest that thrombin that is generated from prothrombin as one of the end results of the activation of the clotting cascade (e.g., at sites of vascular damage) might participate in stimulating SMC proliferation both directly, as a mitogen,[116,117] and indirectly, by stimulating nearly endothelial cells to release PDGFc.

B. Growth Factors from Vascular Smooth Muscle Cells

A second vascular cell that can make a PDGFc under certain conditions is the smooth muscle cell itself. We have recently reported[62] that aortic SMC cultured from the aortas of 2-week-old rats secrete substantial amounts of PDGFc, whereas SMC from adult rats do not. Rates of secretion of PDGFc by the pup rat SMC are comparable to the rates of secretion by vascular endothelial cells. Goat anti-PDGF IgG can abolish the binding of pup SMC PDGFc to test cells and totally eliminates the ability of pup SMC-conditioned medium to stimulate ^{3}H-thymidine incorporation by 3T3 cells. Thus, in contrast to growth factor release from platelets and endothelial cells, PDGFc appears to be the only mitogen secreted by pup rat SMC, at least when assayed on 3T3 test cells.

Since SMC express PDGF receptors and respond mitogenically to PDGF, the PDGFc produced by pup rat SMC may be an example of what Sporn and Todaro[118] have called an "autocrine hormone": a hormone that stimulates the cells that produce it. Two observations support this possibility:

1. Pup rat SMC express fewer cell-surface PDGF receptors than do adult rat SMC.[62] We have shown that the number of PDGF receptors on adult rat SMC after preincubation with pup rat SMC-conditioned medium is reduced toward the level of PDGF receptors expressed by pup rat SMC. This suggests that PDGFc production by pup rat SMC may be sufficient to explain their low levels of PDGF receptors through a mechanism of receptor occupation and downregulation. If this is the case, it is reasonable to propose that the pup rat SMC would be mitogenically stimulated by this PDGFc in an autocrine fashion.
2. If the pup rat SMC were producing a self-mitogen that was not produced by adult rat SMC, we would expect to see a difference in the mitogen requirements of these cells. Indeed, we have found that pup rat SMC do have a higher ^{3}H-thymidine labeling index in low concentrations of PDS than do adult rat SMC.[119]

We have recently reported that adult rat SMC might be activated to secrete PDGFc by conditions associated with wounding in vivo.[120] In these studies, the carotid artery was injured by scraping away the endothelium with an inflated balloon catheter. Two weeks after this injury, extensive proliferation of SMC in the intima of the artery had occurred, and the ^{3}H-thymidine labeling index of the intimal SMC was still high.[31] At this point, the intima was dissected away from the media and cultured. When the PDGF-related phenotype of the intimal SMC was compared with that of SMC from uninjured media, it was found that the number of PDGF receptors expressed by the intimal SMC was only 50% as high as the level of PDGF receptors expressed by cells from the uninjured media. In addition, cells from intima produced up to tenfold higher levels of PDGF than did SMC from the uninjured media. The differences between intimal and medial SMC seem to be at least partially stable in culture since the differences between medial and neointimal smooth cells were observed using cells that had been cultured in vitro for at least 3 weeks. These findings suggest that production of PDGFc by SMC might be activated in response to wounding and might supplement the PDGF derived from attached, degranulated platelets and from activated endothelial cells in contributing to the extensive and prolonged proliferation of SMC in the intima of the wounded vessel (Figure 2). The relative importance of these three sources of PDGF in stimulating SMC proliferation is not yet known and may depend on the type of injury and on the time after injury.

C. Growth Factors from Leukocytes

Interleukin-1 (Il-1) is a member of a group of polypeptide factors which mediate some of the complex series of interactions that occur between the cells of the immune system[121] and which may also affect connective tissue cell types. However, recent data from Libby et al.[90] suggest that whereas Il-1 stimulates proliferation of human fibroblasts, it does not significantly stimulate human, bovine, or rabbit SMC.

Leibovich and Ross[122] observed delayed healing of skin wounds in monocytopenic guinea pigs and reported that culture medium conditioned by macrophages can stimulate growth of cultured guinea pig wound fibroblasts. Liebovich[123] later reported that

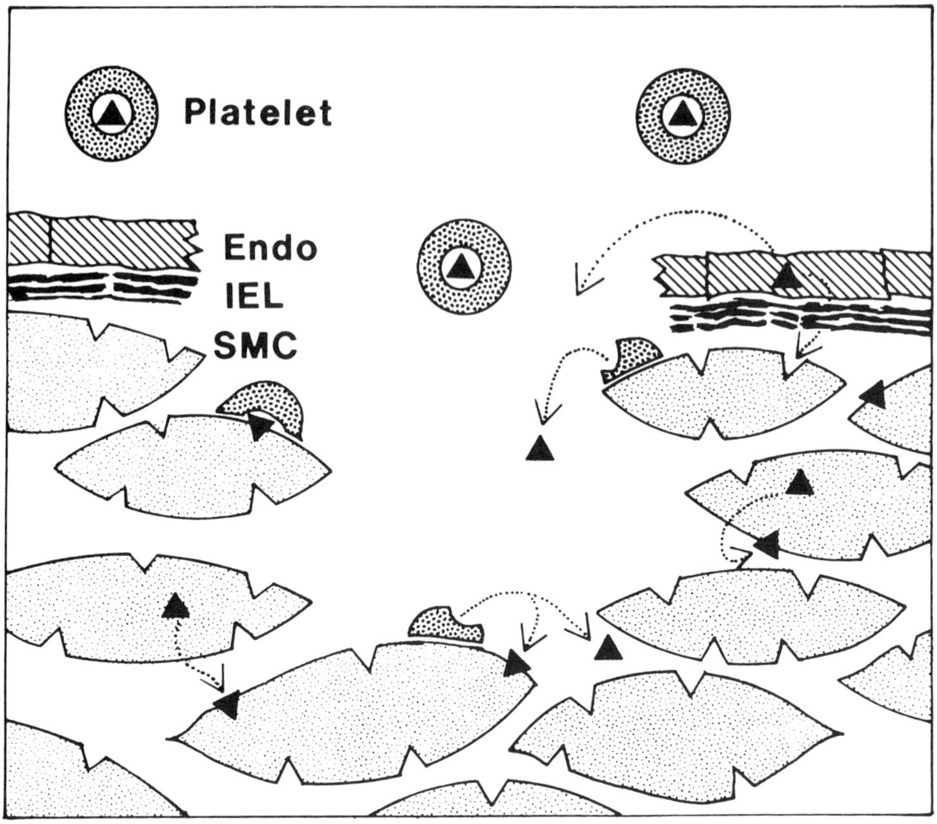

FIGURE 2. Sources of PDGFc for vascular SMC in an injured artery. A small injury to a musculoelastic artery is cartooned as a break in the endothelial cell monolayer (diagonal hatching) and in the internal elastic lamina (IEL, shown as rough parallel bundles) and with some loss of SMC (light stipples). Platelets are dark stippled disks containing PDGF (triangles) within alpha-granules (open circles). The illustration depicts release of PDGFc from adherent platelets, and adjacent endothelial cells and SMC, and its binding to receptors (triangular notches) on SMC. (From Bowen-Pope, D. F. and Seifert, R. A., *Cancer Cells*, Vol. 3, Ferramisco, J., Ozanne, B., and Stiles, C., Eds., Cold Spring Harbor Press, Cold Spring Harbor, N.Y., 1985, 183. With permission.)

the presence of PDS was required for production of growth factor and suggested that the growth factor was produced by modification of a plasma constituent. Subsequent experiments, however, indicate that macrophages can release growth factor activity even in serum-free medium.[124-127] In the case of macrophages derived from peripheral blood monocytes, it seems that serum plays a role in the maturation of the monocytes into macrophages capable of producing growth factor but that, once mature, the serum is not longer necessary.[127] Growth factor production by macrophages is stimulated by attachment to the substratum[128] and further stimulated by several agents, including endotoxin, concanavalin A, opsinized zymosan, and phorbol myristate acetate.[124-127]

The growth factors released by macrophages have not yet been purified or completely characterized. Martin et al.[126] reported that the growth-promoting activity produced by mouse peritoneal macrophages was not adsorbed by carboxymethyl sephadex under conditions which removed PDGF. Shimokado et al.[129] found that human peritoneal and lung macrophages do produce a PDGF-like factor but that most of this factor is present in the form of a complex which does not bind to PDGF receptors in a radioreceptor assay. Complexes may be in equilibrium with some free PDGF since the conditioned medium does contain mitogenic activity which is neutralized by antiserum to PDGF. Molecular seive chromatography under acidic conditions allows the PDGF-like activity to be separated from a PDGF-binding activity. This concomitant production of growth factor and growth factor-binding activities is reminiscent of production of IGF-II and its binding protein by cultured rat liver cells.[130]

Medium conditioned by cultured macrophages obtained from Schilling stainless steel wound chambers implanted subcutaneously in rabbits has been reported to stimulate proliferation of adult bovine aortic SMC.[131] When such macrophages were co-cultured with SMC, the growth rate of the SMC was increased. "Wound fluid" obtained from the Schilling chambers was also stimulatory, possibly due to the presence of growth factors produced by macrophages within the chambers.[131] Mouse peritoneal macrophages produce growth factor activity promoting proliferation and ^{3}H-thymidine in corporation by bovine aortic SMC. The amount of growth factor released was increased by incubation of the macrophages with latex, lipopolysaccharide, or phorbol myristate acetate.[126] Human peripheral monocytes, which have been cultured 5 days in homologous serum to permit maturation into macrophage-like cells, produced considerable growth-promoting activity (^{3}H-thymidine incorporation) for human SMC.[125] Endotoxin and concanavalin A stimulated increased rates of growth factor production.

Given the many pathological conditions in which macrophages are observed in association with proliferating SMC (e.g., Faggiotto et al.[132]) and the evidence mentioned above that cultured macrophages can produce a growth factor(s) active on SMC, it seems likely that growth factors from macrophages play a role in stimulating SMC in the artery wall.

IV. SYMPATHETIC NERVES AND VASOACTIVE PEPTIDES

Recognition of the roles of arterial SMC hypertrophy[3,133] and hyperplasia[134,135] in hypertension provides a rational basis for examining the effects of vasoactive agents on SMC growth. A variety of studies have implicated sympathetic nerves and endogenous catecholamines in both normal growth and development of SMC and in the growth response in hypertension. They suggest that the sympathetic nervous system may exert a chronic form of control of artery wall structure and/or function in addition to its well-known short-term control of vascular tone.[136] Bevan[137] and Bevan and Tsuru[138] found that denervation of the central ear artery of growing rabbits produced decreases in medial thickness and ^{3}H-thymidine-labeled SMC nuclei and increased the sensitivity of the vessels to norepinephrine when compared to contralateral innervated vessels. Rusterholz and Mueller[139] reported a decreased resistance to flow in the maximally dilated rabbit ear vascular bed following unilateral superior cervical ganglionectomy. Cavallero et al.[140] reported that epinephrine injections stimulated DNA synthesis in the aorta and pulmonary artery of the rabbit. Thadani and Schanberg[141] found that alpha adrenergic stimulation induced rat aortic ornithine decarboxylase activity and suggested a possible role for sympathetic nerves in regulation of protein synthesis by vascular SMC. Autonomic nerves may, therefore, exert a trophic influence on artery wall growth and metabolism in a manner resembling that of motor nerves on skeletal muscle.[142] However, for issues relevant to the artery wall, it is difficult to distinguish between direct effects of adrenoceptor stimulation on cell growth and indirect effects due to changes in pressure and wall tension in the intact vasculature. This is particularly true for hypertension. Judy et al.[143] reported that sympathetic nerve activity is significantly greater in spontaneously hypertensive rats (SHR) than in normotensive controls. Antiadrenergic drugs[144] or sympathectomy[145] arrested the onset of hypertension in young SHR and prevented the development of medial hypertrophy and enhanced lysine incorporation into noncollagen protein in medium-sized muscular arteries. Similarly, Rorive et al.[146] found that inhibition of aortic DNA replication in renal hypertensive rats by antihypertensive drugs depended on inhibition of the sympathetic nervous system.

While these studies point to a role for catecholamines in ASMC growth in hypertension, the mechanisms for this effect are unknown. Addition of norepinephrine or epinephrine directly to ASMC in culture results in relatively minor effects on cell growth. Blaes and Boissel[147] reported that epinephrine augments the growth of cultured rat ASMC in a serum-dependent fashion, but the increment in cell numbers was slight (20 to 30%). Similar results were reported for human aortic SMC growth in the presence and absence of the vasoactive peptides angiotensin II and vasopressin.[148] Augmentation of proliferation by these peptides was abolished by passage of the homologous serum employed over CM-Sephadex. While these results do not support a role for catecholamines or vasoactive peptides as complete mitogens for ASMC, they suggest that these agents may act synergistically with serum-derived mitogens to augment rates of cell growth. Whether this augmentation reflects an anabolic effect of cellular metabolism and protein syntheis or represents a more specific action in cell cycle progression is not clear at this time. Nevertheless, if such an effect occurs in vivo, it might suggest at least one way that factors associated with stress and hypertension function as risk factors for atherosclerosis.

V. FIBROBLAST GROWTH FACTOR

Fibroblast growth factor (FGF) is a polypeptide factor, or family of factors, first observed in extracts of bovine brain and pituitary.[149] The molecular identity of all factors with FGF activity is not yet clear. Lemmon et al.[150] have reported that that major activity in pituitary extracts is a basic polypeptide, while brain extracts contain a basic polypeptide (which may be identical to that from the pituitary) and also an acidic polypeptide.

FGF at 100 ng/mℓ has been reported to increase the saturation density of bovine aortic SMC in a base of either 2.5% PDS or WBS.[151] In this system, FGF was considerably more potent than 100 ng/mℓ EGF. When grown in 10% serum, the initial rate of proliferation and final saturation density of first or second passage bovine aortic SMC was less in PDS than in WBS, and growth in PDS could be increased to that in WBS by the addition of 100 ng/mℓ FGF.[71] In serum-free medium supplemented with HDL, FGF stimulated SMC proliferation with an ED_{50} of 5 ng/mℓ and a maximum effect at 100 ng/mℓ.[72] Stimulation by FGF was synergistic with HDL, Somatomedin C, or insulin. Tiell et al.[98] speculated that the absence of FGF might account for the suppressive effect of hypophysectomy on the SMC proliferation seen in response to balloon catheter deendothelialization. To date, however, there is not data as to whether FGF is present in blood or in tissue outside the brain and pituitary.

REFERENCES

1. Thomas, W. A. and Kim, D. N., Atherosclerosis as a hyperplastic and/or neoplastic disease, *Lab. Invest.*, 48, 245, 1983.
2. Ross, R., Wight, T. N., Strandness, E., and Thiele, B., Human atherosclerosis. I. Cell constitution and characteristics of advanced lesions of the superficial femoral artery, *Am. J. Pathol.*, 114, 79, 1984.

3. Owens, G. K., Rabinovitch, P. S., and Schwartz, S. M., Smooth muscle cell hypertrophy versus hyperplasia in hypertension, *Proc. Natl. Acad. Sci. U.S.A.*, 78, 7759, 1981.
4. Chamley-Campbell, J., Campbell, G. R., and Ross, R., The smooth muscle cell in culture, *Physiol. Rev.*, 59, 1, 1979.
5. Jarmolych, J., Dauod, A. S., Landau, J., Fritz, E., and McElvene, E., Aortic media explants. Cell proliferation and production of mucopolysaccharides, collagen, and elastic tissue, *Exp. Mol. Pathol.*, 9, 171, 1968.
6. Fritz, K. E., Jarmolych, J., and Daoud, A. S., Association of DNA synthesis and apparent dedifferentiation of aortic smooth muscle cells *in vitro, Exp. Mol. Pathol.*, 12, 354, 1970.
7. Thyberg, J., Palmberg, L., Nilsson, J., Ksiazek, T., and Sjolund, M., Phenotypc modulation in primary cultures of arterial smooth muscle cells. On the role of platelet-derived growth factor, *Differentiation*, 25, 156, 1983.
8. Paule, W. J., Electron microscopy of the newborn rat aorta, *J. Ultrastruct. Res.*, 8, 219, 1963.
9. Cliff, W. J., The aortic tunica media in growing rats studied with the electron microscope, *Lab. Invest.*, 17, 599, 1967.
10. Gerrity, R. G. and Cliff, W. J., The aortic tunica media of the developing rat. I. Quantitative stereologic and biochemical analysis, *Lab. Invest.*, 32, 585, 1975.
11. Poole, J. C. F., Cromwell, S. B., and Benditt, E. P., Behavior of smooth muscle cells and formation of extracellular structure in the reaction of arterial walls to injury, *Am. J. Pathol.*, 62, 391, 1971.
12. Rhee, C. Y., Herz, F., and Spaet, T. H., Accelerated culture of aortic smooth muscle cells, *Thromb. Res.*, 11, 90, 1977.
13. Grunwald, J. and Haudenschild, C. C., Intimal injury *in vivo* activates vascular smooth muscle cell migration and explant outgrowth *in vitro, Arteriosclerosis*, 4, 183, 1984.
14. Chamley-Campbell, J. H., Campbell, G. R., and Ross, R., Phenotype-dependent response of cultured aortic smooth muscle to serum mitogens, *J. Cell Biol.*, 89, 379, 1981.
15. Wissler, R. W., The arterial medial cell: smooth muscle or multifunctional mesenchyme, *Circulation*, 36, 1, 1967.
16. Moss, N. S. and Benditt, E. P., Spontaneous and experimentally induced arterial lesions. I. An ultrastructural survey of the normal chicken aorta, *Lab. Invest.*, 22, 166, 1970.
17. Schmid, E., Osborn, M., Rungger-Brandle, E., Gabbiani, G., Weber, K., and Franke, W. W., Distribution of vimentin and desmin filaments in smooth muscle tissue of mammalian and avian aorta, *Exp. Cell Res.*, 137, 329, 1982.
18. Carrell, A. and Burrows, M. T., Cultivation of tissues *in vitro* and its technique, *J. Exp. Med.*, 13, 387, 1911.
19. Balk, S. D., Calcium as a regulator of the proliferation of normal, but not of transformed, chicken fibroblasts in a plasma-containing medium, *Proc. Natl. Acad. Sci. U.S.A.*, 68, 271, 1971.
20. Kohler, N. and Lipton, A., Platelets as a source of fibroblast growth-promoting activity, *Exp. Cell Res.*, 87, 297, 1974.
21. Ross, R., Glomset, J., Kariya, B., and Harker, L., Platelet-dependent serum factor that stimulates the proliferation of arterial smooth muscle cells in vitro, *Proc. Natl. Acad. Sci. U.S.A.*, 71, 1207, 1974.
22. Westermark, B. and Wasteson, Å., A platelet factor stimulating human normal glial cells, *Exp. Cell Res.*, 98, 170, 1976.
23. Kaplan, D. R., Chao, F. C., Stiles, C. D., Antoniades, H. N., and Scher, C. D., Platelet alpha granules contain a growth factor for fibroblasts, *Blood*, 53, 1043, 1979.
24. Rutherford, R. B. and Ross, R., Platelet factors stimulate fibroblasts and smooth muscle cells quiescent in plasma serum to proliferate, *J. Cell Biol.*, 69, 196, 1976.
25. Burke, J. M. and Ross, R., Collagen synthesis by monkey arterial smooth muscle cells during proliferation and quiescence in culture, *Exp. Cell Res.*, 107, 387, 1977.
26. Witte, L. D., Kaplan, K. L., Nossel, H. L., Lages, B. A., Weiss, H. J., and Goodman, D. S., Studies of the release from human platelets of the growth factor for cultured human arterial smooth muscle cells, *Circ. Res.*, 42, 402, 1978.
27. Thorgeirsson, G. and Robertson, A. L., Platelet factors and the vascular wall. Variations in growth response between endothelial and medial smooth muscle cells, *Atherosclerosis*, 30, 67, 1978.
28. Ihnatowycz, I. O., Cazenave, J. P., Mustard, J. F., and Moore, S., The effect of a platelet-derived growth factor on the proliferation of rabbit arterial smooth muscle cells in tissue culture, *Thromb. Res.*, 14, 477, 1979.
29. Nilsson, J. and Thyberg, J., Fine structure of arterial smooth muscle cells cultured in the presence of whole blood serum or plasma-derived serum, *Cell Tissue Res.*, 223, 87, 1982.
30. Ross, R., Nist, C., Kariya, B., Rivest, M. J., Raines, E., and Callis, J., Physiological quiescence in plasma-derived serum: influence of platelet-derived growth factor on cell growth in culture, *J. Cell. Physiol.*, 97, 497, 1978.

31. Clowes, A. W., Reidy, M. A., and Clowes, M. M., Kinetics of cellular proliferation after arterial injury. I. Smooth muscle growth in the absence of endothelium, *Lab. Invest.*, 59, 327, 1983.
32. Vogel, A., Raines, E., Kariya, B., Rivest, M. J., and Ross, R., Coordinate control of 3T3 cell proliferation by platelet-derived growth factor and plasma components, *Proc. Natl. Acad. Sci. U.S.A.*, 75, 2810, 1978.
33. Singh, J. P., Chaikin, M. A., and Stiles, C. D., Phylogenetic analysis of platelet-derived growth factor by radioreceptor assay, *J. Cell Biol.*, 95, 667, 1982.
34. Huang, J. S., Huang, S. S., and Deuel, T. F., Human platelet-derived growth factor: radioimmunoassay and discovery of a specific plasma-binding protein, *J. Cell Biol.*, 97, 383, 1983.
35. Bowen-Pope, D. F., Malpass, T. W., Foster, D. M., and Ross, R., Platelet-derived growth factor *in vivo*: levels, activity, and rate of clearance, *Blood*, 64, 458, 1984.
36. Libby, P. and O'Brien, K. V., Culture of quiescent arterial smooth muscle cells in a defined serum-free medium, *J. Cell. Physiol.*, 115, 217, 1983.
37. Roberts, A. B., Anzano, M. A., Lamb, L. C., Smith, J. M., and Sporn, M. B., New class of transforming growth factors potentiated by epidermal growth factor: isolation from non-neoplastic tissues, *Proc. Natl. Acad. Sci. U.S.A.*, 78, 5339, 1981.
38. Assoian, R. K., Grotendorst, G. R., Miller, D. M., and Sporn, M. B., Cellular transformation by coordinated action of three peptide growth factors from human platelets, *Nature (London)*, 309, 804, 1984.
39. Childs, C. B., Proper, J. A., Tucker, R. F., and Moses, H. L., Serum contains a platelet-derived transforming growth factor, *Proc. Natl. Acad. Sci. U.S.A.*, 79, 5312, 1982.
40. Assoian, R. K., Komoriya, A., Meyers, C. A., Miller, D. M., and Sporn, M. B., Transforming growth factor-β in human platelets. Identification of a major storage site, purification, and characterization, *J. Biol. Chem.*, 258, 7155, 1983.
41. Antoniades, H. N., Scher, C. D., and Stiles, C. D., Purification of human platelet-derived growth factor, *Proc. Natl. Acad. Sci. U.S.A.*, 76, 1809, 1979.
42. Heldin, C.-H., Westermark, B., and Wasteson, Å., Platelet-derived growth factor: purification and partial characterization, *Proc. Natl. Acad. Sci. U.S.A.*, 76, 3722, 1979.
43. Deuel, T. F., Huang, J. S., Proffitt, R. T., Baenziger, J. U., Chang, D., and Kennedy, B. B., Human platelet-derived growth factor — purification and resolution into two active protein fractions, *J. Biol. Chem.*, 256, 8896, 1981.
44. Raines, E. and Ross, R., Platelet-derived growth factor. I. High yield purification and evidence for multiple forms, *J. Biol. Chem.*, 257, 5154, 1982.
45. Antoniades, H. N. and Scher, C. D., Radioimmunoassay of a human serum growth factor for Balb/c 3T3 cells: derivation from platelets, *Proc. Natl. Acad. Sci. U.S.A.*, 74, 1973, 1977.
46. Raines, E. W., unpublished data, 1985.
47. Bowen-Pope, D. F. and Ross, R., Platelet-derived growth factor. II. Specific binding to cultured cells, *J. Biol. Chem.*, 257, 5161, 1982.
48. Heldin, C.-H., Westermark, B., and Wasteson, Å., Demonstration of an antibody against platelet-derived growth factor, *Exp. Cell Res.*, 136, 255, 1981a.
49. Nilsson, J., Ksiazek, T., Heldin, C.-H., and Thyberg, J., Demonstration of stimulatory effects of platelet-derived growth factor on cultivated rat arterial smooth muscle cells, *Exp. Cell Res.*, 145, 231, 1983b.
50. Temin, H. M., Pierson, R. W., Jr., and Dulak, N. C., The role of serum in the control of multiplication of avian and mammalian cells in culture, in *Growth, Nutrition and Metabolism of Cells in Culture*, Vol. 5, Rothblat, G. H. and Cristofalo, V. J., Eds., Academic Press, New York, 1972, 50.
51. Williams, L. T., Tremble, P., and Antoniades, H. N., Platelet-derived growth factor binds specifically to receptors on vascular smooth muscle cells and the binding becomes nondissociable, *Proc. Natl. Acad. Sci. U.S.A.*, 79, 5867, 1982.
52. Weinstein, R., Stemerman, M. B., and Maciag, T., Hormonal requirement for growth of arterial smooth muscle cells in vitro: an endocrine approach to atherosclerosis, *Science*, 818, 1981.
53. Nilsson, J. and Olsson, A. G., Prostaglandin E_1 inhibits DNA synthesis in arterial smooth muscle cells stimulated with platelet-derived growth factor, *Atherosclerosis*, 53, 77, 1984.
54. Atkinson, P. R., Weidman, E. R., Bhaumick, B., and Bala, R. M., Release of somatomedin-like activity by cultured WI-38 human fibroblasts, *Endocrinology*, 106, 2006, 1980.
55. Clemmons, D. R., Multiple hormones stimulate the production of somatomedin by cultured human fibroblasts, *J. Clin. Endocrinol. Metab.*, 58, 850, 1984.
56. Pietila, K., Moilanen, T., and Nikkari, T., Prostaglandins enhance the synthesis and inhibit the growth of rabbit aortic smooth muscle cells in culture, *Artery*, 7, 509, 1980.
57. Huttner, J. J., Gwebu, E. T., Panganamala, R. V., Milo, G. E., and Cornwell, D. G., Fatty acids and their prostaglandin derivatives — inhibitors of proliferation in aortic smooth muscle cells, *Science*, 197, 289, 1977.

58. Heldin, C.-H., Westermark, B., and Wasteson, Å., Specific receptors for platelet-derived growth factor on cells derived from connective tissue and glia, *Proc. Natl. Acad. Sci. U.S.A.*, 78, 3664, 1981b.
59. Huang, J. S., Huang, S. S., Kennedy, B., and Deuel, T. F., Platelet-derived growth factor: specific binding to target cells, *J. Biol. Chem.*, 257, 8130, 1982.
60. Bowen-Pope, D. F., Seifert, R. A., and Ross, R., The platelet-derived growth factor receptor, in *Cell Proliferation: Recent Advances*, Boynton, A. L. and Leffert, H., Eds., Academic Press, New York, 1985, 287.
61. Glenn, K. C., Bowen-Pope, D. F., and Ross, R., Platelet-derived growth factor. III. Identification of a platelet-derived growth factor receptor by affinity labeling, *J. Biol. Chem.*, 257, 5172, 1982.
62. Seifert, R. A., Schwartz, S. M., and Bowen-Pope, D. F., Developmentally regulated production of platelet-derived growth factor-like molecules, *Nature (London)*, 311, 1984.
63. DiCorleto, P. E. and Bowen-Pope, D. F., Cultured endothelial cells produce a platelet-derived growth factor-like protein, *Proc. Natl. Acad. Sci. U.S.A.*, 80, 1919, 1983.
64. Huang, J. S., Huang, S. S., and Deuel, T. F., Specific covalent binding of platelet-derived growth factor to human plasma $alpha_2$-macroglobulin, *Proc. Natl. Acad. Sci. U.S.A.*, 81, 342, 1984.
65. Raines, E. W., Bowen-Pope, D. F., and Ross, R., Plasma binding proteins for platelet-derived growth factor that inhibit its binding to cell-surface receptors, *Proc. Natl. Acad. Sci. U.S.A.*, 81, 3424, 1984.
66. Adamson, E. D. and Rees, A. R., Epidermal growth factor receptors, *Mol. Cell. Biochem.*, 34, 129, 1981.
67. Savage, C. R. and Cohen, S. J., Epidermal growth factor and a new derivative. Rapid isolation procedures and biological and chemical characterization, *J. Biol. Chem.*, 247, 7609, 1972.
68. Byyny, R. L., Orth, D. N., Cohen, S., and Doyne, E. S., Epidermal growth factor: effects of androgens and andrenergic agents, *Endocrinology*, 95, 776, 1974.
69. Bowen-Pope, D. F. and Ross, R., Is epidermal growth factor present in human blood? Alteration of EGF binding specificity in the radioreceptor assay, *Biochem. Biophys. Res. Commun.*, 114, 1036, 1983.
70. Oka, Y. and Orth, D. N., Human plasma epidermal growth factor/Beta-urogastrone is associated with blood platelets, *J. Clin. Invest.*, 72, 249, 1983.
71. Gospodarowicz, D. and Ill, C. R., Do plasma and serum have different abilities to promote cell growth? *Proc. Natl. Acad. Sci. U.S.A.*, 77, 2726, 1980.
72. Gospodarowicz, D., Hirabayashi, K., Giguere, L., and Tauber, J. P., Factors controlling the proliferative rate, final cell density, and life span of bovine vascular smooth cells in culture, *J. Cell Biol.*, 89, 568, 1981.
73. Bhargava, G., Rifas, L., and Makman, M. H., Presence of epidermal growth factor receptors and influence of epidermal growth factor on proliferation and aging in cultured smooth muscle cells, *J. Cell. Physiol.*, 100, 365, 1979.
74. Sporn, M. B., Newton, D. L., Roberts, A. B., DeLarco, J. E., and Todaro, G. J., Retinoids and suppression of the effects of polypeptide transforming factors — a new molecular approach to chemoprevention of cancer, in *Molecular Actions and Targets for Cancer Chemotherapeutic Agents*, Sartorelli, A. C., Lazo, J. S., and Bertino, J. R., Eds., Academic Press, New York, 1981, 541.
75. Macpherson, I. and Montagnier, L., Agar suspension culture for the selective assay of cells transformed by polyoma virus, *Virology*, 23, 291, 1968.
76. Todaro, G. J., DeLarco, J. E., and Cohen, S., Transformation by murine and feline sarcoma viruses specifically blocks binding of epidermal growth factor to cells, *Nature (London)*, 264, 26, 1976.
77. Marquardt, H., Hunkapiller, M. W., Hood, L. E., Twardzik, D. R., DeLarco, J. E., Stephenson, J. R., and Todaro, G. J., Transforming growth factors produced by retrovirus-transformed rodent fibroblasts and human melanoma cells: amino acid sequence homology with epidermal growth factor, *Proc. Natl. Acad. Sci. U.S.A.*, 80, 4684, 1983.
78. Bowen-Pope, D. F., Vogel, A., and Ross, R., Production of platelet-derived growth factor-like molecules and reduced expression of platelet-derived growth factor receptors accompanies transformation by a wide spectrum of agents, *Proc. Natl. Acad. Sci. U.S.A.*, 81, 2396, 1984.
79. Anzano, M. A., Roberts, A. B., DeLarco, J. E., Wakefield, L. M., Assoian, R. K., Roche, N. S., Smith, J. M., Lazarus, J. E., and Sporn, M. B., Increased secretion of type B transforming growth factor accompanies viral transformation of cells, *Mol. Cell. Biol.*, 5, in press.
80. Gerrard, J. M., Phillips, D. R., Rao, G. H. R., Plow, E. F., Walz, D. A., Ross, R., Harker, L. A., and White, J. G., Biochemical studies of two patients with the gray platelet syndrome, *J. Clin. Invest.*, 66, 102, 1980.
81. Weiss, H. J., Platelet physiology and abnormalities of platelet function, *N. Engl. J. Med.*, 293, 531, 1975.

82. Baumgartner, H. R. and Muggli, R., Adhesion and aggregation: morphological demonstration and quantitation in vivo and in vitro, in *Platelets in Biology and Pathology,* Gordon, J. I., Ed., Elsevier North-Holland Biomedical Press, New York, 1976, 23.
83. Goldberg, I. D., Stemerman, M. B., and Handin, R. I., Vascular permeation of platelet factor 4 after endothelial injury, *Science,* 209, 611, 1980.
84. Harker, L. A., Ross, R., Slichter, S. J., and Scott, C. R., Homocystine-induced arteriosclerosis: the role of endothelial cell injury and platelet response in its genesis, *J. Clin. Invest.,* 58, 731, 1976.
85. Moore, S., Friedman, R. J., Singal, D. P., Gauldie, J., Blajchman, M. A., and Roberts, R. S., Inhibition of injury induced thromboatherosclerotic lesions by antiplatelet serum in rabbits, *Thromb. Haemost.,* 35, 70, 1976.
86. Fuster, V. and Bowie, E. J. W., The von Willebrand pig as a model for atherosclerotic research, *Thromb. Haemost.,* 39, 322, 1978.
87. Friedman, R. J., Stemerman, M. B., Wenz, B., Moore, S., Gauldie, J., Gent, M., Tiell, M. L., and Spaet, T. H., The effect of thrombocytopenia on experimental atherosclerotic lesion formation in rabbits. Smooth muscle cell proliferation and reendothelialization, *J. Clin. Invest.,* 60, 1191, 1977.
88. Blundell, T. L. and Humbel, R. E., Hormone families: pancreatic hormones and homologous growth factors, *Nature (London),* 287, 781, 1980.
89. Zapf, J., Froesch, E. R., and Humbel, R. E., The insulin-like growth factors (IGF) of human serum: chemical and biological characterization and aspects of their possible physiological role, *Curr. Top. Cell. Regul.,* 19, 257, 1981.
90. Libby, P., Wyler, D. J., Janicka, M. W., and Dinarello, C. A., Differential effects of human interleukin-1 on growth of human fibroblasts and vascular smooth muscle and endothelial cells, ms. submitted.
91. Taggart, H. and Stout, R. W., Control of DNA synthesis in cultured vascular endothelial and smooth muscle cells, *Atherosclerosis,* 37, 549, 1980.
92. Pfeifle, B., Ditschuneit, H. H., and Ditschuneit, H., Insulin as a cellular growth regulator of rat arterial smooth muscle cells in vito, *Horm. Metab. Res.,* 12, 381, 1980.
93. Pfeile, B. and Ditschuneit, H., Effect of insulin on growth of cultured human arterial smooth muscle cells, *Diabetologia,* 20, 155, 1981.
94. Koschinsky, T., Bunting, C. E., Schwippert, B., and Gries, A., Regulation of diabetic serum growth factors for human muscular cells by the metabolic control of diabetes mellitus, *Atherosclerosis,* 39, 313, 1981.
95. Ledet, T., Growth hormone stimulating the growth of arterial medial cells in vitro. Absence of effect of insulin, *Diabetes,* 25, 1011, 1976.
96. Ledet, T., Fischer-Dzoga, K., and Wissler, R. W., Growth of rabbit aortic smooth muscle cells cultured in media containing diabetic and hyperlipemic serum, *Diabetes,* 25, 207, 1976.
97. Ledet, T., Growth hormone antiserum suppresses the growth effect of diabetic serum. Studies on rabbit aortic medial cell cultures, *Diabetes,* 26, 798, 1977.
98. Tiell, M. L., Stemmerman, M. B., and Spaet, T. H., The influence of the pituitary on arterial intimal proliferation in the rat, *Circ. Res.,* 42, 644, 1978.
99. Bettmann, M. A., Stemerman, M. B., and Ransil, B. J., The effect of hypophysectomy on experimental endothelial cell regrowth and intimal thickening in the rat, *Circ. Res.,* 48, 907, 1981.
100. Fischer-Dzoga, K., Fraser, R., and Wissler, R. W., Stimulation of proliferation in stationary primary cultures of monkey and rabbit aortic smooth muscle cells. I. Effect of lipoprotein fractions of hyperlipemic serum and lymph, *Exp. Mol. Pathol.,* 24, 346, 1975.
101. Fischer-Dzoga, K. and Wissler, R. W., Stimulation of proliferation in stationary primary cultures of monkey aortic smooth muscle cells. II. Effect of varying concentrations of hyperlipemic serum and low density lipoproteins of varying dietary fat origins, *Atherosclerosis,* 24, 515, 1976.
102. Fless, G. M., Kirchhausen, T., Fischer-Dzoga, K., Wissler, R. W., and Scanu, A. M., Serum low density lipoproteins with mitogenic effect on cultured aortic smooth muscle cells, *Atherosclerosis,* 41, 171, 1982.
103. Brown, B. G., Mahley, R., and Assmann, G., Swine aortic smooth muscle in tissue culture. Some effects of purified swine lipoproteins on cell growth and morphology, *Circ. Res.,* 39, 415, 1976.
104. Pietila, K., Long-term effect of hyperlipidemic serm on the synthesis of glycosaminoglycans and on the rate of growth of rabbit aortic smooth muscle cells in culture, *Atherosclerosis,* 42, 67, 1982.
105. Ross, R. and Glomset, J. A., Atherosclerosis and the arterial smooth muscle cells, *Science,* 180, 1332, 1973.
106. Hessler, J. R., Robertson, A. L., and Chisolm, G. M., LDL-induced cytoxicity and its inhibition by HDL in human vascular smooth muscle and endothelial cells in culture, *Atherosclerosis,* 32, 213, 1979.
107. Schwartz, E., Bienkowski, R. S., Coltoff-Schiller, B., Goldfischer, S., and Blumenfeld, O. O., Changes in the components of extracellular matrix and in growth properties of cultured aortic smooth muscle cells upon ascorbate feeding, *J. Cell Biol.,* 462, 1982.

108. Gajdusek, C., DiCorleto, P., Ross, R., and Schwartz, S. M., An endothelial cell-derived growth factor, *J. Cell Biol.,* 85, 467, 1980.
109. Wang, C.-H., Largis, E. E., and Schaffer, S. A., The effects of endothelial cell-conditioned media on the proliferation of aortic smooth muscle cells and 3T3 cells in culture, *Artery,* 9, 358, 1981.
110. Castellot, J. J., Addonizio, M. L., Rosenberg, R., and Karnovsky, M. J., Cultured endothelial cells produce a heparin-like inhibitor of smooth muscle cell growth, *J. Cell Biol.,* 90, 372, 1981.
111. DiCorleto, P. E., Cultured endothelial cells produce multiple growth factors for connective tissue cells, *Exp. Cell Res.,* 153, 167, 1984.
112. Clowes, A. W. and Karnovsky, M. J., Suppression by heparin of smooth muscle cell proliferation in injured arteries, *Nature (London),* 265, 625, 1977.
113. Barrett, T. B., Gajdusek, C. M., Schwartz, S. M., McDougall, J. K., and Benditt, E. P., Expression of the *sis* gene by endothelial cells in culture and *in vivo, Proc. Natl. Acad. Sci. U.S.A.,* 81, 6772, 1984.
114. Fox, P. L. and DiCorleto, P. E., Regulation of production of a platelet-derived growth factor-like protein by cultured bovine aortic endothelial cells, *J. Cell. Physiol.,* 121, 298, 1984.
115. Harlan, J., Thompson, P., Ross, R., and Bowen-Pope, D. F., Alpha-thrombin induces release of PDGF-like molecules by cultured human umbilical endothelial cells, *J. Cell Biol.,* in press.
116. Chen, L. B. and Buchanan, J. M., Mitogenic activity of blood components. I. Thrombin and prothrombin, *Proc. Natl. Acad. Sci. U.S.A.,* 72, 131, 1975.
117. Carney, D. H., Glenn, K. C., and Cunningham, D. D., Conditions which affect initiation of animal cell division by trypsin and thrombin, *J. Cell. Physiol.,* 95, 13, 1978.
118. Sporn, M. B. and Todaro, G. J., Autocrine secretion and malignant transformation of cells, *N. Engl. J. Med.,* 303, 878, 1980.
119. Bowen-Pope, D. F. and Seifert, R. A., Exogenous and endogenous sources of PDGF-like molecules and their potential roles in vascular biology, in *Cancer Cells,* Vol. 3, Ferramisco, J., Ozanne, B., and Stiles, C., Eds., Cold Spring Harbor Press, Cold Spring Harbor, N.Y., 1985, 185.
120. Walker, L. N., Bowen-Pope, D. F., and Reidy, M. A., Secretion of platelet-derived growth factor (PDGF)-like activity by arterial smooth muscle cells in induced as a response to injury, *J. Cell Biol.,* 99 (Abstr.), 416a, 1984.
121. Dinarello, C. A., Interleukin-1, *Rev. Infect. Dis.,* 6, 51, 1984.
122. Leibovich, S. J. and Ross, R., A macrophage-dependent factor that stimulates the prevention of fibroblasts *in vitro, Am. J. Pathol.,* 84, 501, 1976.
123. Leibovich, S. J., Production of macrophage-dependent fibroblast stimulating activity (M-FSA) by murine macrophages. Effects of BALB/c 3T3 fibroblasts, *Exp. Cell Res.,* 113, 47, 1978.
124. Wahl, S. M., Wahl, L. M., McCarthy, J. B., Chedid, L., and Mergenhagen, S. E., Macrophage activation by mycobacterial water soluble compounds and synthetic muramyl dipeptide, *J. Immunol.,* 122, 2226, 1979.
125. Glenn, K. C. and Ross, R., Human monocyte-derived growth factor(s) for mesenchymal cells: activation of secretion by endotoxin and con A, *Cell,* 25, 603, 1981.
126. Martin, B. M., Gimbrone, M. A., Unanue, E. R., and Cotran, R. S., Stimulation of non-lymphoid mesenchymal cell proliferation by a macrophage-derived growth factor, *J. Immunol.,* 126, 1510, 1981.
127. Leslie, C. C., Musson, R. A., and Henson, P. M., Production of growth factor activity for fibroblasts by human monocyte-derived macrophages, *J. Leuk. Biol.,* 36, 143, 1984.
128. Bitterman, P. B., Rennard, S. I., Hunninghake, G. W., and Crystal, R. G., Human alveolar macrophage growth factor for fibroblasts, *J. Clin. Invest.,* 70, 806, 1982.
129. Shimokado, K., Raines, E. W., Madtes, D. K., Barrett, T. B., Benditt, E. P., and Ross, R., A significant part of macrophage-derived growth factor consists of at least two forms of PDGF, *Cell,* 43, 277, 1986.
130. Moses, A. C., Nissley, S. P., Passamini, J., and White, R. M., Further characterization of growth hormone-dependent somatomedin binding proteins in rat serum and demonstration of somatomedin-binding proteins produced by rat liver cells in culture, *Endocrinology,* 104, 536, 1979.
131. Greenburg, G. and Hunt, T., The proliferative response *in vivo* of vascular endothelial and smooth muscle cells exposed to wound fluids and macrophages, *J. Cell. Physiol.,* 97, 333, 1978.
132. Faggiotto, A., Ross, R., and Harker, L., Studies of hypercholesterolemia in the nonhuman primate. I. Changes that lead to fatty streak formation, *Arteriosclerosis,* 4, 323, 1984.
133. Olivetti, G., Melissari, M., Marchetti, G., and Anversa, P., Quantitative structural changes of the rat thoracic aorta in early spontaneous hypertension: tissue composition, and hypertrophy and hyperplasia of smooth muscle cells, *Circ. Res.,* 51, 19, 1982.
134. Fernandez, D. and Crane, W. A. J., New cell formation in rats with accelerated hypertension due to partial aortic constriction, *J. Pathol.,* 100, 307, 1970.
135. Bevan, R. D., An autoradiographic and pathological study of cellular proliferation in rabbit arteries correlated with an increase in arterial pressure, *Blood Vessels,* 13, 100, 1976.

136. Bevan, R. D. and Tsuru, H., Long-term denervation of vascular smooth muscle causes not only functional but structural changes, *Blood Vessels,* 16, 109, 1979.
137. Bevan, R. D., Effect of sympathetic denervation on smooth muscle cell proliferation in growing rabbit ear artery, *Circ. Res.,* 37, 14, 1975.
138. Bevan, R. D. and Tsuru, H., Functional and structural changes in the rabbit ear artery after sympathetic denervation, *Circ. Res.,* 49, 478, 1981.
139. Rusterholz, D. B. and Mueller, S. M., Sympathetic nerves exert a chronic influence on the intact vasculature that is age related, *Ann. Neurol.,* 11, 365, 1982.
140. Cavallero, C., DiTondo, U., Mingazzini, P. L., Pesando, P. C., and Spagnoli, L. G., Cell proliferation in the atherosclerotic lesions of cholesterol-fed rabbits. II. Histological, ultrastructural, and radioautographic observations on epinephrine-treated rabbits, *Atherosclerosis,* 17, 49, 1973.
141. Thadani, P. V. and Schanberg, S. M., Effect of stress and sympathetic activity on rat cardiac and aortic ornithine decarboxylase activity, *Life Sci.,* 25, 1009, 1979.
142. Gutmann, E., Neutrophic relations, *Annu. Rev. Physiol.,* 38, 177, 1976.
143. Judy, W. V., Watanabe, A. M., Henry, D. P., Besch, H. R., Murphy, W. R., and Hockel, G. M., Sympathetic nerve activity. Role in regulation of blood pressure in the spontaneously hypertensive rat, *Circ. Res.,* 38 (Suppl. 2), 21, 1976.
144. Nakada, T. and Lovenberg, W., Lysine incorporation in vessels of spontaneously hypertensive rats: effects of adrenergic drugs, *Eur. J. Pharmacol.,* 48, 87, 1978.
145. Yamori, Y., Nakada, T., and Lovenberg, W., Effect of antihypertensive therapy on lysine incorporation into vascular protein of the spontaneously hypertensive rat, *Eur. J. Pharmacol.,* 38, 349, 1976.
146. Rorive, G. L., Carlier, P. J., and Foidart, J. M., Hyperplasia of rat arteries smooth muscle cells associated with development and reversal of renal hypertension, *Clin. Sci.,* 59, 335S, 1980.
147. Blaes, N. and Boissel, J. P., Growth-stimulating effect of catecholamines on rat aortic smooth muscle cells in culture, *J. Cell. Physiol.,* 116, 167, 1983.
148. Campbell-Boswell, M. and Robertson, A. L., Jr., Effects of angiotensin II and vasopressin on human smooth muscle cells *in vitro, Exp. Mol. Pathol.,* 35, 265, 1981.
149. Gospodarowicz, D., Localization of a fibroblast growth factor and its effect alone and with hydrocortisone on 3T3 cell growth, *Nature (London),* 249, 123, 1974.
150. Lemmon, S. K., Riley, M. C., Thomas, K. A., Hoover, G. A., Maciag, T., and Bradshaw, R., Bovine fibroblast growth factor: comparison of brain and pituitary preparations, *J. Cell Biol.,* 95, 162, 1982.
151. Gospodarowicz, D., Moran, J., and Braun, D., Control of proliferation of bovine vascular cells, *J. Cell. Physiol.,* 91, 377, 1977.
152. Bowen-Pope, D. F., The growth factor-platelet-transformation connection, in *The Cell Surface in Development and Cancer,* a volume of *Developmental Biology,* Browder, L., Ed., Plenum Press, New York, in press.

Chapter 7

HEPARIN AND THE REGULATION OF GROWTH IN THE VASCULAR WALL

John J. Castellot, Jr. and Morris J. Karnovsky

TABLE OF CONTENTS

I. INTRODUCTION

A salient characteristic of the normal arterial wall is that the component cells are in a quiescent growth state.[1] The importance of the intimal endothelial cells (EC) which line the lumen in maintaining this state has been recognized for nearly a decade, but the precise molecular mechanisms underlying this phenomenon are only beginning to be clarified.

When EC are damaged, a series of events ensues which leads to the migration and proliferation of the medial smooth muscle cells (SMC). First, platelets adhere to the injured area and release their granular contents. One of the α-granule products to which the SMC are exposed is the platelet-derived growth factor (PDGF), which is both chemotactic and mitogenic for SMC.[2,3] Although the vast majority of platelets are present only transiently at the injured site, there are several other sources of SMC mitogens which are present. These include macrophages[4,5] and the regenerating EC.[6] One of the currently intriguing hypotheses of the pathogenesis of arteriosclerosis — the response to injury theory — holds that SMC respond to the presence of the platelet, macrophage, and EC products by migrating into the intima and proliferating, thus giving rise to a hyperplastic mass.[7-11] Uninjured EC at the periphery of the injury migrate and proliferate until the damaged area is re-endothelialized, at which time SMC growth ceases.[7] Repeated cycles of injury and proliferation may then lead to full-blown lesions with their potentially serious consequences.

Many laboratories have provided experimental support for the response to injury hypothesis, including our own. Much of this work has been directed at understanding the proliferative response of SMC to various mitogenic substances and is detailed elsewhere in this volume. Our work has focused on the growth regulatory mechanisms within the vascular wall in both physiological and pathological conditions. In particular, we have studied the potential role of endogenous heparin-like species as regulators of SMC growth.

II. HISTORICAL PERSPECTIVE

A. In Vivo Models for EC Injury Link SMC Proliferation to EC Regeneration

In the mid-1970s, we and others began to examine histologically the response of arteries to an EC injury. The air-dry injury method developed in this laboratory involved ligating a 1.5-cm segment of the rat carotid artery, evacuating the contents, and passing a gentle stream of air through the lumen under carefully controlled pressure, humidity, and time.[7] The endothelium is desiccated and begins to peel off within minutes after restoring blood flow to the carotid. Platelets then carpet the exposed intima, and a dramatic proliferation of SMC is observed within the next 7 to 14 days. The advantages to the air-dry procedure is that the injury to the vessel wall is well controlled, destroying all of the EC without significant injury to the media. The rat carotid artery eliminates complications in studying endothelial regeneration, since there were no ostia in the segment used, and EC regeneration occurs solely from the uninjured proximal and distal portions of the vessel. SMC proliferation, myointimal thickening, and EC regeneration can be quantitated by morphometric and autoradiographic means.

When this type of injury was produced (Figure 1), we observed little myointimal proliferation at the proximal and distal regions of the injury in which re-endothelialization was rapid. In the middle of the segment, however, where endothelial regeneration occurred last, there was massive SMC hyperplasia. The proliferation reached a peak at 10 to 14 days after injury, at which time re-endothelialization was complete.

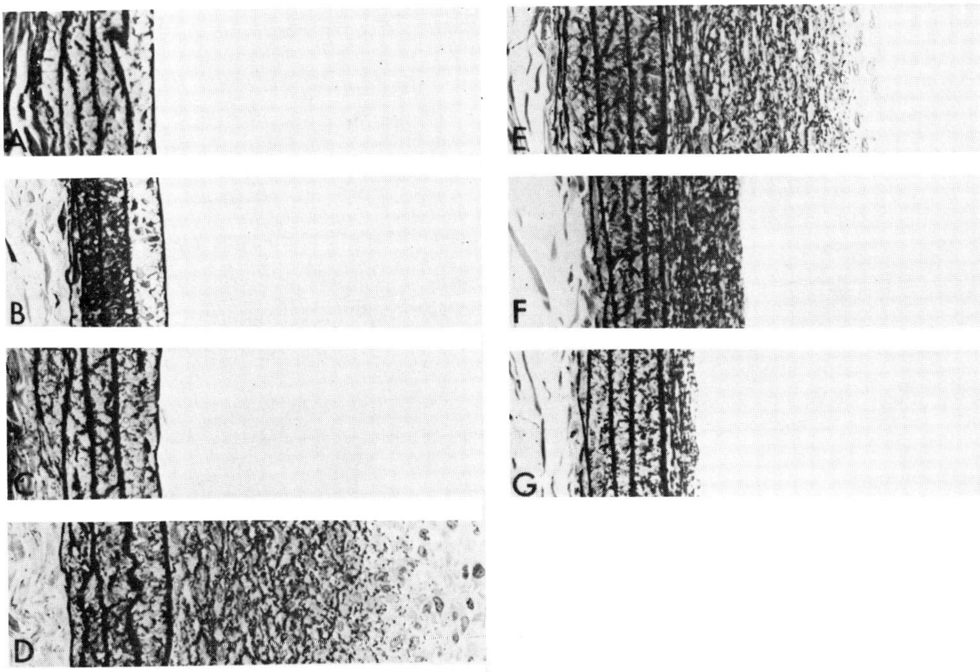

FIGURE 1. Histology of the arterial wall in response to EC injury. Composite histologic figure illustrating the changes occurring in the wall of the carotid artery after endothelial denudation. To aid comparison of intimal thickness at each stage, the elastic laminae of the media are aligned. The vessel lumen is at the right of each picture. This procedure stains EC rather lightly; thus they show up only faintly when photographed. (A) Normal vessel with flat endothelium; (B) central area of the segment 4 days after denudation, with loss of endothelium and showing prominent edema of the innermost zone of the media; (C) newly covered area 7 days after denudation, with plump EC and showing resolution of the medial edema; (D) central area of the segment 14 days after denudation, showing the massive extent of myointimal thickening; (E) central area at 17 days, showing a slight decrease in the thickness of the myointimal layer; (F) and (G) central area at 1 and 3 months, respectively, showing progressive regression of the thickened zone.

In many cases, considerable regression of the hyperplastic lesion had occurred by 1 to 3 months after the injury. Similar results have been obtained by others[11,12] in the aorta, using a balloon catheter-induced injury. This was the first clue we had that EC and/or other cells might produce growth effectors for SMC.

B. Heparin Blocks the Proliferative Response of SMC In Vivo

At about the time that the air-dry injury model was developed, Chen and Buchanan[13] reported that thrombin was mitogenic for certain cell types. Because the clotting sequence is activated in vascular injury, we tested the effect of heparin on the SMC hyperplasia in the rat carotid after the air-dry injury.[11] Heparin was continuously infused into control and injured animals via a catheter placed in the left jugular vein, with the tip lying in the superior vena cava. With this treatment, the SMC proliferation usually seen after EC injury was almost completely abolished (Figure 2). We found no effects of heparin on the number of circulating or deposited platelets or on their structure and degranulation. Neither the rate nor the extent of re-endothelialization was affected by heparin.

Although heparin had markedly suppressed myointimal proliferation, the original rationale — that the antiproliferative effect would be mediated by the antithrombin, anticoagulant activity of heparin — remained an unanswered question. With the sep-

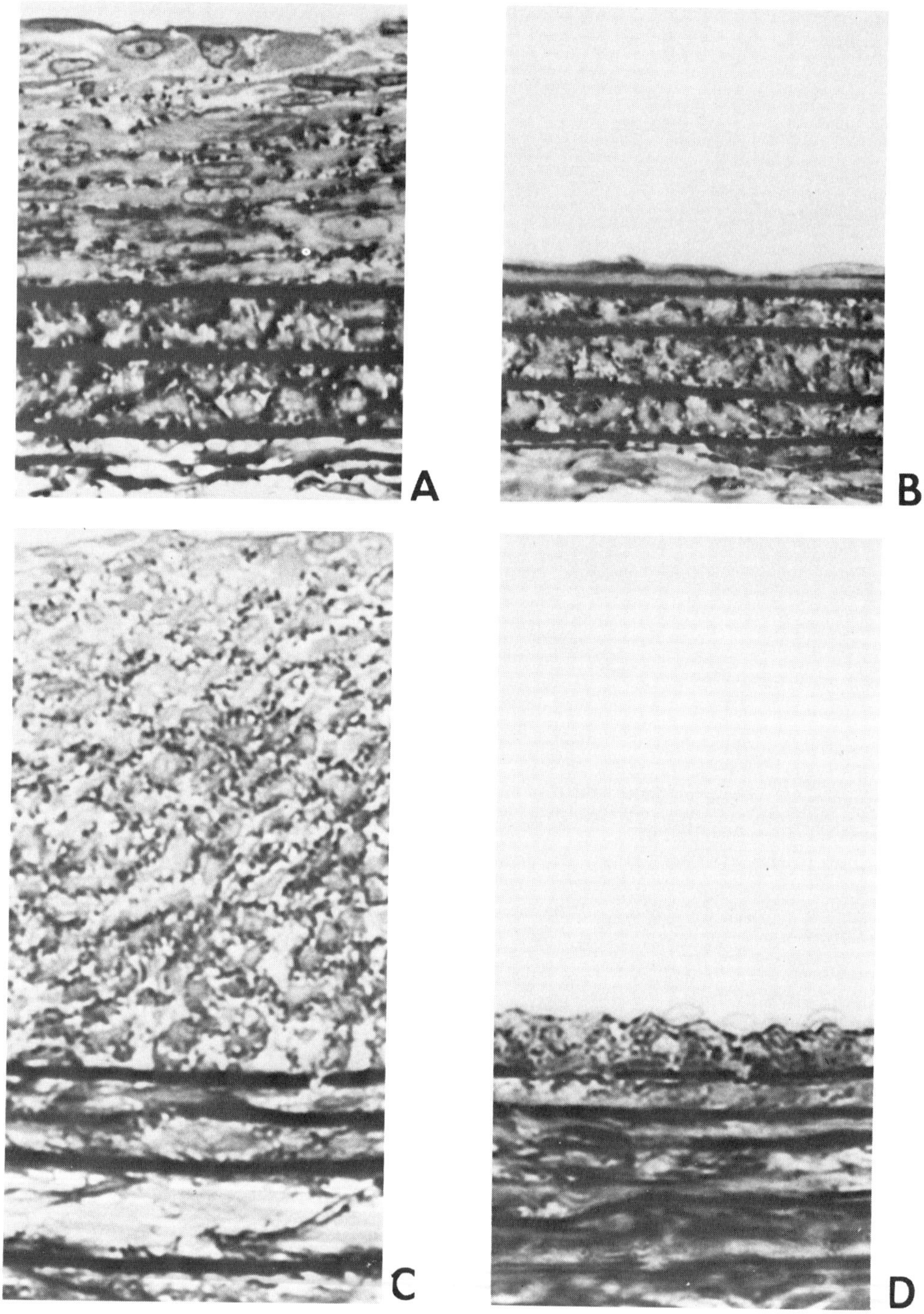

FIGURE 2. Effect of AC and NAC heparin on SMC proliferation following EC injury in vivo. Histological sections are shown of the central segments of air-dried rat carotid arteries 14 days after injury. (A) and (C) Control rats not treated with heparin — note the substantial myointimal proliferation in these animals; (B) rat treated with AC heparin (approximately 0.5 mg/kg/hr) in the same experiment as the rat in panel A — note the greatly reduced degree of myointimal thickening; (D) rat treated with NAC heparin (approximately 0.5 mg/kg/hr) in the same experiment as the rat in panel C — again, note the greatly reduced degree of myointimal thickening.

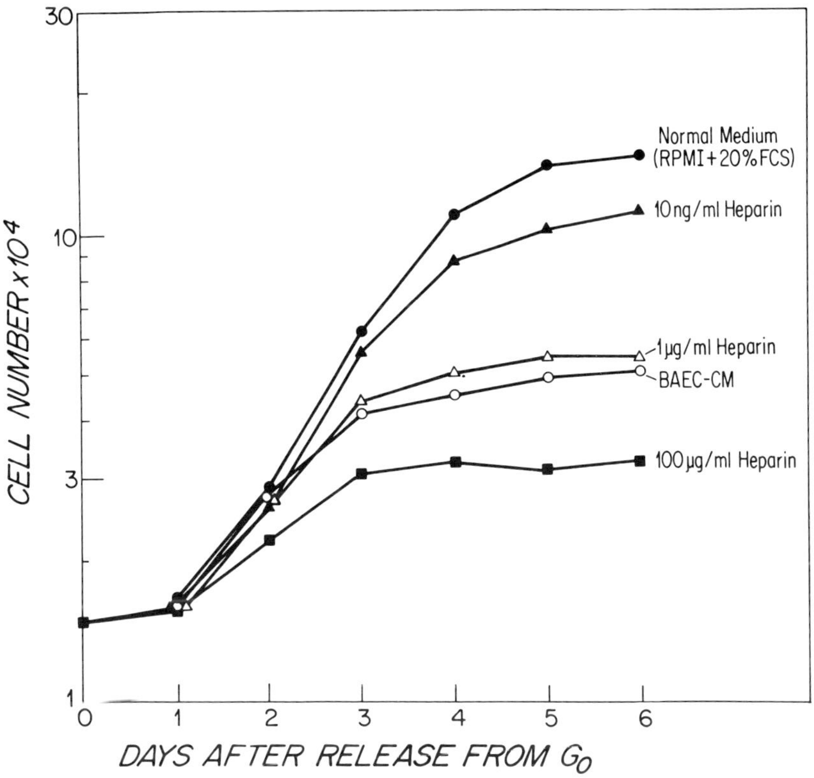

FIGURE 3. Effect of heparin and EC-conditioned medium on SMC growth. Sparsely plated rat aortic SMC were growth-arrested by serum deprivation in G_0. Control cultures were released from the G_0 block by placing them in their normal growth medium (RPMI + 20% fetal calf serum). These cells proliferate until they are confluent. Other cultures were placed in a 1:1 mixture of bovine aortic EC-conditioned medium and RPMI + fetal calf serum (final concentration, 20% fetal calf serum; BAEC-CM) or in RPMI + 20% fetal calf serum containing the indicated concentrations of heparin. Cell numbers were measured at daily intervals. The cells were not fed during the experiment. (From Castellot, J. J., Rosenberg, R. D., and Karnovsky, M. J., *Biology of Endothelial Cells*, Jaffe, E. A., Ed., Martinus Nijhoff, Boston, 1984, 118. With permission.)

aration of heparin into purified anticoagulant (AC) and nonanticoagulant (NAC) fractions by Rosenberg's group[15] and by Lindahl and his colleagues,[16] it now became possible to answer this question directly. In collaboration with Dr. Rosenberg and using his NAC heparin fraction, we repeated the infusion experiments.[17] It was clear that on a mass basis, NAC heparin was as effective as AC material in blocking SMC proliferation (Figure 2). Thus, the original rationale for trying heparin was wrong, and the discovery of the antiproliferative activity of heparin was quite serendipitous. We switched our attention to cultured SMC to characterize the antiproliferative activity of heparin.

C. Cultured SMC Are More Responsive to Heparin than Other Cell Types

To test the ability of heparin to inhibit the growth of cultured SMC, we first growth-arrested the cells at subconfluent densities by serum deprivation. This was done to more closely mimic the in vivo situation in which SMC normally exist in a quiescent growth state. The SMC were released from the growth-arrested (i.e., G_0) state by replacing the serum-poor medium with normal growth medium (containing 20% serum)

Table 1
SENSITIVITY OF VARIOUS CELL TYPES TO HEPARIN

Cell type	ED_{50} (μg/mℓ)
Rat aortic SMC	2
Rat mesenteric artery SMC	7
Calf aortic SMC	4
Cebus monkey aorta SMC	3
Human umbilical vein SMC	4
Canine tracheal SMC	80
Rat kidney glomerular mesangial cells	5
Baby hamster kidney fibroblasts (BHK)	500
Polyoma virus-transformed BHK	>800[a]
Swiss 3T3	>600[a]
BALB/c 3T3	500
Chinese hamster embryo fibroblasts (normal)	500
Chinese hamster embryo fibroblasts (transformed)	>600[a]
Madin-Darby canine kidney epithelial cells	200
Rat kidney epithelial cells	200
Bovine aortic endothelial cells	400
Rat epididymal endothelial cells	300
Mouse epididymal endothelial cells	200
Human umbilical vein endothelial cells	500

Note: The above cell types were growth-arrested and then exposed to a wide range of heparin concentrations (10 ng/mℓ to 1 mg/mℓ). Growth inhibition was determined by cell counts.

[a] An accurate ED_{50} could not be determined because cytotoxicity occurred above that concentration.

in the presence or absence of heparin. The antiproliferative effect was determined by counting cell number at various intervals after release from growth arrest. Heparin was very effective in suppressing the growth of cultured SMC[18,19] (Figure 3). The ED_{50} for the antiproliferative activity is in the 1- to 10-μg/mℓ range.

Many other cell types were examined for their responsiveness to heparin[20] (Table 1). Under these assay conditions, only SMC and related cell types (e.g., mesangial cells) were sensitive to heparin. NAC and AC heparins were found to be equally antiproliferative. There are numerous reports in the literature on the antiproliferative effect of heparin on various nonvascular cells, mostly fibroblasts (cf. Lippman and Mathews[21]). The dose required for 50% inhibition was very high, usually in the 1-mg/mℓ range, compared with 1- to 10-μg/mℓ doses needed for SMC. It should be pointed out that in these experiments, heparin was added to exponentially growing cultures. An interesting characteristic of our system is that SMC which are quiescent (i.e., in G_0) before their exposure to growth medium containing heparin are 50 to 100 times more sensitive to heparin than SMC which are exponentially growing at the time they are exposed. The possible explanations for this will be discussed later.

D. Other Glycosaminoglycans Do Not Inhibit SMC Growth

We used our inhibition assay to compare the effects of heparin and other classes of glycosaminoglycans as SMC antiproliferative agents.[19] Whereas AC and NAC heparins were highly potent and effective (50% inhibition at 1 to 10 μg/mℓ), other glycosaminoglycans such as dermatan sulfate, chondroitin sulfates A and C, and hyaluronic acid required at least 500 μg/mℓ doses to inhibit SMC growth by 50%.

E. EC Secrete a Heparin-Like Inhibitor of SMC Proliferation

In the air-dry model, the cessation of SMC hyperplasia in a particular region of the damaged segment was correlated with the re-endothelialization of that area, as discussed above. We therefore thought it possible that there was a regulatory influence of EC on SMC proliferation and undertook to examine this using cultured cells. Conditioned medium from confluent cultures of EC exerted a strong inhibitory effect on SMC released from G_0 (Figure 3).[19] Interestingly, exponentially growing EC produced a stimulatory effect on SMC growth, thus suggesting that cessation of EC growth after regeneration in vivo may be an important step in the regulation and inhibition of SMC proliferation. The mitogenic factor elaborated by EC has been explored in detail by others.[6]

Biochemical characterization[19] (Table 2) showed that the antiproliferative activity from EC was unaffected by heating and was also insensitive to proteases, hyaluronidases, and chondroitin sulfate ABC lyase. However, a highly purified Flavobacterium heparinase almost completely abolished the activity. This enzyme, which has no detectable activity against proteins or other glycosaminoglycans, degrades heparin into disaccharide and smaller fragments that lack inhibitory activity, as discussed later. In addition, a trichloroacetic acid-soluble, ethanol-insoluble glycosaminoglycan fraction prepared from EC-conditioned medium was highly antiproliferative. When this material was treated with the bacterial heparinase, the activity was abolished. Therefore, we concluded[19] that the inhibitory activity produced by EC under certain requisite conditions was a heparin-like molecule. The possibility that mast cell contamination of the primary cultures was contributing to the presence of antiproliferative heparin in the conditioned medium was ruled out by exhaustive cytochemical examination of the cultures.

We found that only SMC were sensitive to the inhibitory activity of EC-conditioned medium. In addition, media conditioned by a variety of other cell types had no effect on SMC proliferation. Thus, there appeared to be both target and producer cell specificity for the antiproliferative effect. Although we originally reported that conditioned medium from SMC did not affect SMC growth, recent work by Rosenberg and colleagues suggests that confluent, but not exponentially growing, SMC synthesize a highly antiproliferative heparin species.[22] This raises the interesting possibility of an autocrine regulatory mechanism operating in SMC that is mediated by heparin. The roles of EC- and SMC-derived heparin is an open question; one possibility is that there is an EC-SMC interaction in which EC-derived heparin acts as a "trigger" for SMC to begin synthesizing their own heparin.

The combination of in vivo and in vitro experiments described above provided a potential explanation for the findings of Rosenberg and Lindahl that more than half of the heparin molecules present in a typical preparation from beef lung or porcine intestine were devoid of anticoagulant activity and also suggested a new biological role for heparin as a growth regulator in the vascular wall. These studies also indicate a pharmacological rationale for the use of heparin that is discussed later.

III. STRUCTURE FUNCTION STUDIES

A. The Minimum Effective Size of Antiproliferative Heparin is a Pentasaccharide

In collaboration with Rosenberg, we carried out experiments to establish what size molecule of heparin was effective in inhibiting SMC growth. Fragments of heparin were prepared by limit digest with nitrous acid, and discrete oligosaccharide fragments were separated by gel filtration chromatography.[23] The smallest heparin fragment with antiproliferative activity was a hexamer, whereas di- and tetrasaccharide fractions were inactive (Table 3). The maximum antiproliferative activity was obtained with dodeca-

Table 2
BIOCHEMICAL CHARACTERIZATION OF THE EC-DERIVED ANTIPROLIFERATIVE ACTIVITY

Treatment	Percent inhibition
None	68 ± 12
60°C, 30 min	63 ± 10
90°C, 5 min	62 ± 5
Trypsin (30 μg/mℓ, 2 h, 37°C)	60 ± 10
Chymotrypsin (30 μg/mℓ, 2 h, 37°C)	64 ± 8
Subtilisin (30 μg/mℓ, 60 min, 37°C)	57 ± 7
Hyaluronidase (15 U/mℓ, 90 min, 37°C)	60 ± 12
Chondroitin 80 ABC lyase (1 U/mℓ, 90 min, 37°C)	65 ± 6
Flavobacterium heparinase (10 U/mℓ, 90 min, 37°C)	15 ± 7

Note: Conditioned medium from confluent, primary bovine aortic EC was subjected to the indicated treatments and then assayed for its ability to inhibit rat SMC. The data is expressed as "percent inhibition". This number is obtained by first subtracting the starting cell from the cell number at the endpoint of the experiment (usually 5 to 7 days), when growth has essentially ceased in both control and conditioned medium-exposed cultures. The net growth is set at 100%, and the net growth in conditioned medium-treated cells is expressed as a percentage of the control. The difference is the percent inhibition.

mer and larger fragments — activities equivalent to native (i.e., commercial) heparin at equimolar doses.

The above experiments were very useful in defining the basic size parameters for the antiproliferative effect. However, the limitation of this approach is that the size fractions obtained by gel filtration, while relatively homogeneous with respect to the size and degree of sulfation, are quite heterogeneous with respect to the specific chemical structures within a size class. This problem has been successfully eliminated by Choay and colleagues, who undertook a massive project to synthesize heparin oligosaccharides from precisely defined monosaccharide units, thus yielding molecules of unambiguous structure.[24] After a 200-step process, Choay's laboratory had synthesized pentasaccharide heparin chains. In collaboration with this group, we tested several of these synthetic heparins in our SMC inhibition assay. A pentasaccharide molecule was found to have approximately the same antiproliferative activity as the hexasaccharide fraction obtained from gel filtration (Table 3). All the tetrasaccharides tested were almost completely inactive.[71]

Heparin interacts with several hemostatic enzymes. It appears that to maximally accelerate the inactivation of thrombin, factor IXa, and factor XIa by antithrombin III, a heparin molecule of at least 16 residues is required, although smaller fragments did have some ability to accelerate the factor Xa-antithrombin III reaction.[25] Thus it

Table 3
COMPARISON OF THE ANTIPROLIFERATIVE CAPACITY, ANTICOAGULANT ACTIVITY, AND DEGREE OF SULFATION OF HEPARIN DERIVATIVES

	Antiproliferative activity[a]			
	1 μg/mℓ	100 μg/mℓ	Anticoagulant activity[b]	Total sulfates[c]
Heparin (60 saccharides)	36	75	205	2.7
Disaccharide	0	4	0	1.8
Tetrasaccharide	2	12	0	2.3
Hexasaccharide	21	46	0	2.7
Octasaccharide	29	65	0	2.7
Decasaccharide	40	71	0	2.7
Dodecasaccharide	43	82	30	2.7
Choay pentasaccharide (with 3-O-sulfate)	18	47	0	3.2
Choay pentasaccharide (without 3-O-sulfate)	9	22	0	2.8
N-desulfated	10	18	0	1.4
N-desulfated; N-resulfated	41	82	62	2.3
N-desulfated; N-reacety-lated	24	64	0	1.0
Totally desulfated	7	10	0	0
Totally desulfated; N-resul-fated	15	26	0	1.0
Totally desulfated; N-re-acetylated	3	12	0	0

[a] Percent inhibition of rat aortic SMC.
[b] USP units of thrombin inhibitory activity per milligram.
[c] Moles SO_4/moles glucosamine.

seems that for complete anticoagulant activity, heparin fragments of 16 residues or greater are needed, whereas maximum antiproliferative activity requires only a 12-saccharide heparin molecule.

B. Both N- and O-Sulfates Are Required for Antiproliferative Activity

To examine which chemical moieties were necessary for the antiproliferative effect, native heparin was chemically modified[26] (Table 3). Totally desulfated heparin lost all its inhibitory activity. The N-desulfated heparin lost most of its antiproliferative activity, as well as all its anticoagulant activity. Acetylation of the N-positions of the N-desulfated molecule restored much of the inhibitory potency, but did not restore the anticoagulant capacity. Because it is not possible chemically to directly O-desulfate heparin, this modification was approximated by first totally desulfating and then re-N-sulfating heparin. This derivative also lost most of its inhibitory potential.

These experiments indicated that O-sulfation, or at least a negative charge at the O position, is necessary for the antiproliferative activity. The situation at the N-position is somewhat different. When the N-sulfates were removed, the N-position is positively charged, and the growth of SMC is not suppressed. Acetylation results in a neutral charge at the N-positions and restored the antiproliferative activity, suggesting that a negative charge at these positions is not required.

To more critically determine the precise molecular determinants of the antiproliferative effect, we have examined the ability of several synthetic heparins provided by

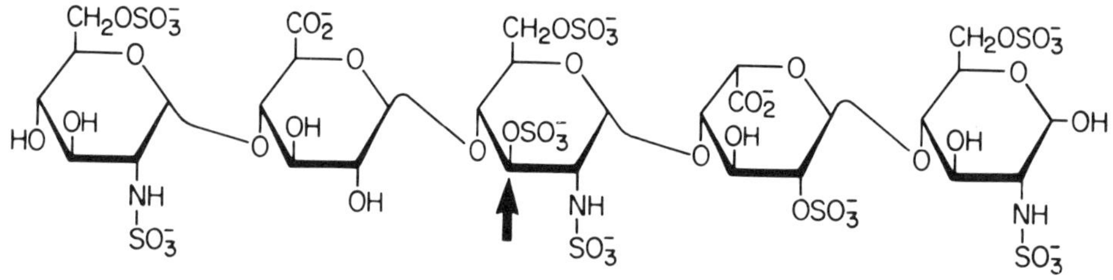

FIGURE 4. Structure of the Choay heparin pentasaccharide. Removal of the 3-O-sulfate denoted by the arrow results in almost complete loss of the antiproliferative activity (Table 3).

Choay to inhibit SMC proliferation. As discussed above, a pentasaccharide species was found to retain significant antiproliferative activity. When a pentasaccharide was synthesized lacking the 3-*O*-sulfate from the internal glucosamine residue, essentially all the antiproliferative activity was lost (Table 3; Figure 4), thus suggesting a critical role for this moiety.

The structure-function studies have a dual purpose: (1) to determine the molecular determinants of the capacity of heparin to inhibit SMC proliferation and (2) to provide molecules which have lost their anticoagulant ability but retain their antiproliferative effect. Such molecules may be potentially useful, as discussed later. Ideally, one would like to isolate heparin from EC or from EC-conditioned medium and chemically analyze it. However, the mass of heparin which can be obtained from either cultured or in vivo endothelium is quite small, and thus this approach has proven to be technically rather difficult. When the endogenous heparin species have been characterized, we speculate that the antiproliferative material from EC and SMC will be rich in the pentasaccharide sequences containing the 3-*O*-sulfate described above.

IV. BINDING AND INTERNALIZATION OF HEPARIN

A. Heparin Binds to Specific Sites on the SMC Surface

To begin to understand how heparin exerts its antiproliferative effect, we carried out studies on the binding and mode of internalization of heparin in heparin-sensitive cell types.[27] The binding of heparin to SMC, utilizing ^{125}I- and ^{3}H-heparin probes, is characterized by specific, saturable binding sites on the cell surface. The affinity ($K_d = 10^{-9}$ *M*) and number of binding sites (100,000 per cell) are typical of regulatory molecules with specific cell surface receptors.

Growth-arrested SMC bound eight times more heparin than exponentially growing cells. We had previously found that growth-arrested SMC were 50 to 100 times more sensitive to the growth inhibitory effect of heparin than exponential cultures;[19] part of this sensitivity may be due to the increased amount of heparin bound. One difference between proliferating and quiescent cells is that quiescent cells have much more cell surface fibronectin, a molecule which has heparin binding sites.[28] We cannot biochemically rule out the possibility of a heparin-fibronectin binding interaction as the reason for the increased heparin binding to quiescent SMC. However, fibronectin is present on the growth-arrested SMC surface in linear arrays, as revealed by indirect immunofluorescence (not shown), whereas heparin displays a diffuse binding pattern (Figure 5), thus making the possibility of a heparin-fibronectin interaction less likely as the explanation for the increased binding of heparin to quiescent SMC. In addition, the reduction in binding was due to the presence of fewer binding sites rather than to a

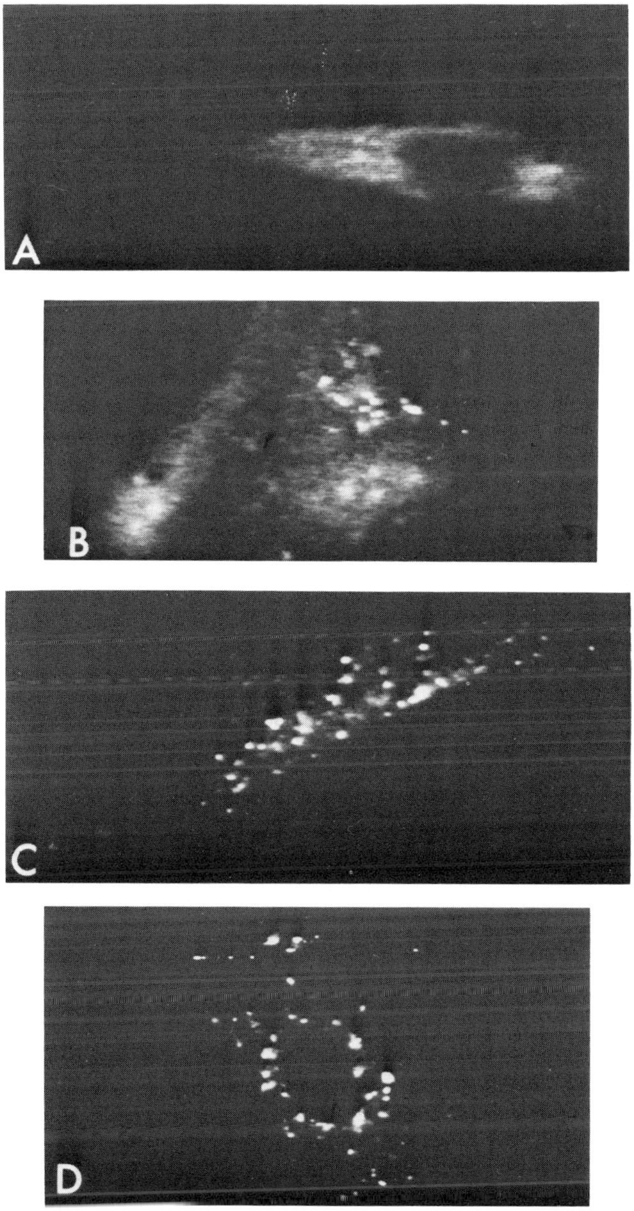

FIGURE 5. Binding and uptake of FITC-heparin. Growth-arrested SMC were exposed to 2 μg/mℓ FITC-heparin for 15 min at 4°C. The FITC-heparin was removed and the cells were placed in normal growth medium at 37°C for the times indicated below. They were then fixed and examined immediately using video image intensification fluorescence microscopy. (A) SMC fixed immediately at 4°C; (B) 2 min at 37°C; (C) 15 min at 37°C; (D) 2 hr at 37°C.

decreased affinity for the ligand, suggesting that a change in receptor number rather than the addition of a new component of binding (such as fibronectin) is responsible.

B. Heparin is Internalized by a Receptor-Mediated Endocytic Mechanism

Work from many laboratories has suggested that cells retrieve biologically important molecules from their environment by characteristic endocytic mechanisms. One of these modes of uptake which recently has been intensely studied is receptor-mediated endocytosis. Several types of ligands including α-2-macroglobulin,[29] low-density lipoprotein,[30] epidermal growth factor,[31] transferrin,[32] Semliki Forest virus,[33] and immunoglobulins[34] appear to follow this pathway, which functions in the delivery of ligands to the endosomal compartment.

We have used two approaches to study the mechanism of heparin uptake and its distribution in SMC. Utilizing FITC-heparin in conjunction with video image intensification fluorescence microscopy (Figure 5), we observed that FITC-heparin was bound to the cell surface at 4°C. Upon warming the cells to 37°C, the FITC-heparin formed surface clusters and was rapidly internalized in endocytic vesicles. These vesicles eventually became localized to the perinuclear region.

Analysis of radiolabeled heparin uptake revealed two distinct rates of uptake (Figure 6). There was an initial rapid entry of labeled heparin into SMC followed by a slow rate of uptake that occurred over several days. The $t_{1/2}$ for the initial rapid uptake phase was approximately 15 min. This is slower than the $t_{1/2}$ of 7 min observed for some receptor-mediated uptake systems,[30] but is comparable to the uptake rate for epidermal growth factor by hepatocytes in the perfused rat liver[35] and faster than the uptake of human chorionic gonadotropin in cultured granulosa cells.[36] The slow phase of uptake could be due to one or more of the following mechanisms: (1) nonfluid phase, nonreceptor-mediated (adsorptive) endocytosis; (2) fluid phase endocytosis, which would require that some of the bound heparin be slowly shed into the medium where it could be internalized by pinocytosis; and (3) receptor-mediated uptake which occurs by more than one rate-limiting mechanism. With regard to this last possibility, it is interesting to note that the biphasic uptake rate is observed even during continuous incubation of the cells with the ligand. This suggests that the rapid receptor-mediated uptake mechanism is saturated very quickly and then is down-regulated or otherwise rendered incompetent for days. The slow phase could then be due to a completely separate uptake mechanism, as just mentioned. Alternatively, the slow phase may represent the rate of reutilization or insertion of new functional receptors into the membrane.

The presence of high-affinity heparin binding sites on the SMC surface, combined with the rapid initial internalization and the pattern of FITC-heparin uptake, strongly suggests that this glycosaminoglycan can enter SMC via receptor-mediated endocytosis. Several other reports on the uptake of glycosaminoglycans by cultured cells exist.[37-39] However, in most cases heparin uptake was not studied, and quantitative measurements of both binding and uptake were not made. In addition, these studies were carried out before receptor-mediated endocytosis was recognized; thus the role of this important internalization mechanism was not examined.

In preliminary experiments, we have observed heparin degradation products in SMC at both early (1 hr) and late (18 hr) time points. However, the techniques for measurement of heparin degradation are not as straightforward as the methods for measuring protein degradation, and more work is needed to clarify this important point.

To our knowledge, these studies provide the first evidence for receptor-mediated endocytosis of a glycosaminoglycan. The radiolabeled and fluorescent heparin probes should prove very useful in answering a number of important questions, including the metabolic fate of the internalized heparin, degradation and recycling of the heparin receptors, and the relationship between the binding and uptake of heparin and its antiproliferative activity.

V. EFFECTS OF HEPARIN ON SMC METABOLISM

A. Heparin Blocks SMC in Late G_1 Phase or at the G_1/S Boundary

Flow microfluorimetric analysis of cell cycle distribution of heparin-treated SMC[20]

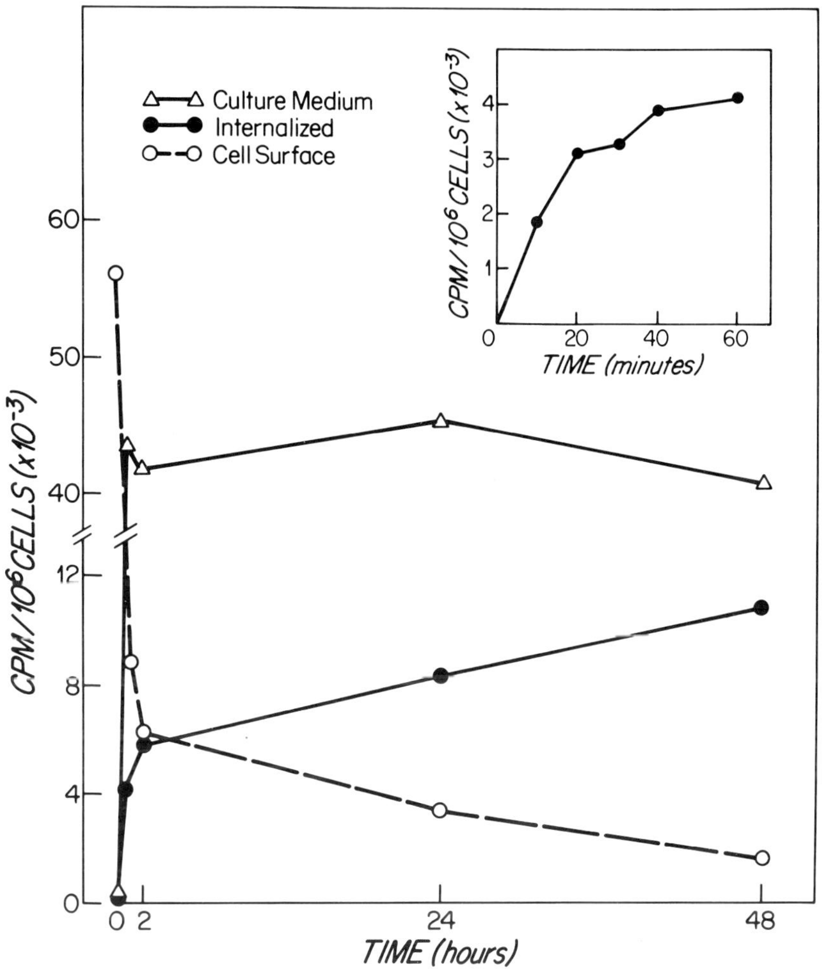

FIGURE 6. Uptake of radiolabeled heparin by SMC. Growth-arrested SMC were incubated in ^{3}H-heparin (13 μg/mℓ, 2×10^6 CPM per sample) for 60 min at 4°C. The cells were washed five times in ice-cold buffer and then warmed to 37°C. At the indicated times, the supernatant (Δ) was harvested and processed for scintillation counting as described. The cells were washed three times and then incubated in a 100-fold excess of unlabeled heparin for 60 min. This was followed by exposure to trypsin (50 μg/mℓ for 15 min). Heparin-displaceable and trypsin-releasable counts were pooled to give total surface counts (○). The cells (•) were solubilized in sodium hydroxide and processed for scintillation counting. The counts obtained from this sample were considered to be inside the cell.

revealed that heparin blocked either the $G_0 \rightarrow S$ transit or a very early S phase event in SMC (Figure 7). The $G_0 \rightarrow S$ transit can be functionally divided into two phases: (1) the competence phase, in which factors such as PDGF stimulate cells to leave the G_0 state and become "competent" to traverse the rest of G_1; (2) the progression phase, in which competent cells move through the rest of G_1 and enter S phase.[40] The progression phase requires a good supply of nutrients such as amino acids and appears to be mediated by factors such as somatomedins and steroids. In addition to the effect of heparin on the $G_0 \rightarrow S$ transition, a small but consistent increase in the G_2 compartment was observed in heparin-treated cells. Whether this represents true G_2 cells or tetraploid G_1 cells is not known.

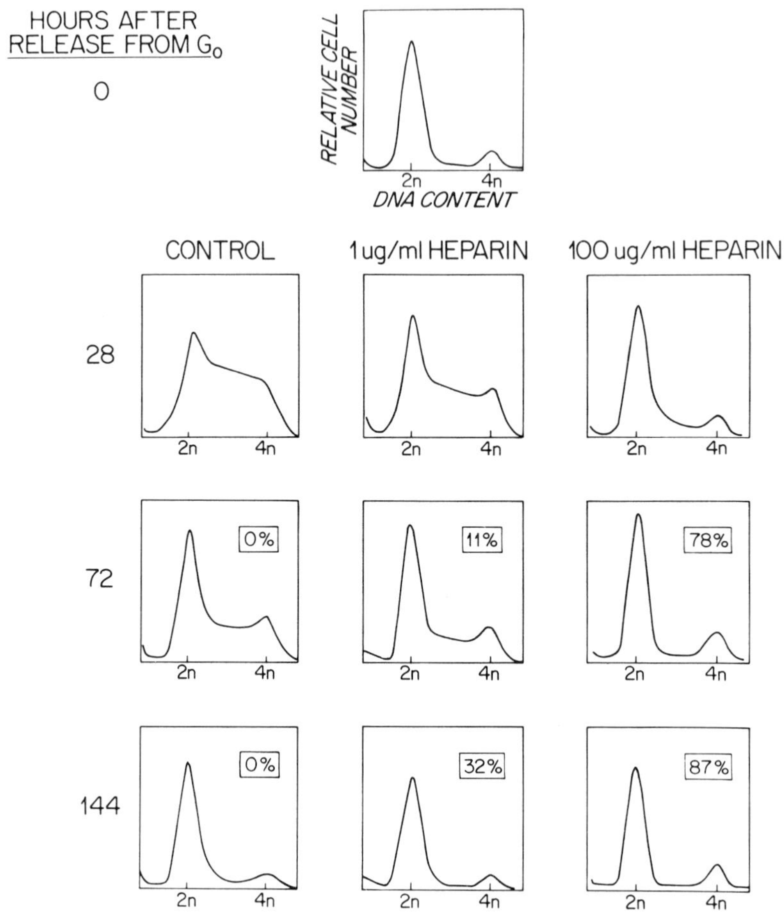

FIGURE 7. Flow microfluorimetric cell cycle analysis of heparin-treated SMC. Rat aortic SMC were growth-arrested and then exposed to normal growth medium with or without the indicated concentrations of heparin. At the indicated intervals, the cells were harvested and processed for flow microfluorimetry. Replicate cultures were harvested for cell counting, and the percent inhibition of growth was determined. DNA histograms are displayed for selected time points to demonstrate the relationship between the different cell cycle compartments and the antiproliferative effect of heparin. Note that at high concentrations of heparin, very few cells leave the growth-arrested state and reenter the cell cycle.

PDGF is a strongly cationic polypeptide that can bind avidly to anionic molecules like heparin. Because PDGF is a potent mitogen for SMC, it was important to determine whether heparin blocked the $G_0 \rightarrow S$ transition via an interaction with PDGF. We therefore determined how long after release from G_0 heparin could be added and still inhibit SMC growth (Figure 8). Heparin could be added up to 18 hr after serum stimulation with little or no loss of its antiproliferative effect. At 24 hr and later time points, however, heparin rapidly lost its ability to block SMC growth. The effectiveness of heparin, even when added 18 hr after serum stimulation, suggests strongly that it does not affect the competence phase. This makes the possibility that heparin inhibits growth via a direct interaction with PDGF highly unlikely, especially since this mitogen has been shown to exert its growth-promoting effect within 60 min after its addition to cells.[40] The loss of the antiproliferative effect after 18 hr of serum stimulation coincides

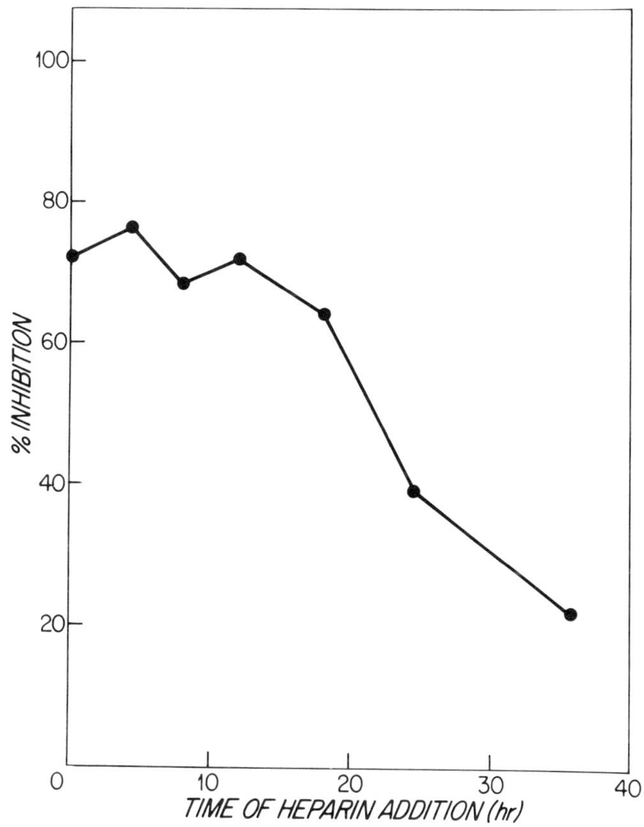

FIGURE 8. Kinetic analysis of heparin on the $G_0 \rightarrow S$ transit. Growth-arrested rat aortic SMC were released from the block by placing them in normal growth medium. At the indicated times after release from G_0, heparin was added to some cultures (final concentration, 100 μg/mℓ). Cell number was determined 7 days after heparin addition. Time points longer than 36 hr were not used because cell division occurs after this time, making interpretation of growth inhibition data difficult.

with the time that the cells are beginning to enter S-phase and provides additional evidence for the hypothesis that heparin is an antiprogression factor.

B. Heparin Inhibits DNA and RNA Synthesis, but not Protein Synthesis, Probably via Blocking Uptake of the Nucleoside Precursors

We determined the effect of heparin on DNA, RNA, and protein synthesis using standard metabolic labeling techniques.[20] SMC were growth-arrested and then released from the G_0 block by placing them in normal growth medium with or without heparin. At various intervals after release from G_0, cells were pulsed for 60 min with the appropriate radiolabeled precursor. The rates of macromolecular synthesis were determined by the incorporation of label into trichloroacetic acid-insoluble material. Immediate effects of heparin on DNA and RNA synthesis were observed, while protein synthesis was unaffected. Both DNA and RNA synthesis were inhibited by 60 to 80% within minutes after adding heparin.

Because of this rapid inhibition of nucleic acid synthesis, we considered the possibility that heparin could be blocking synthesis of these macromolecules by inhibiting

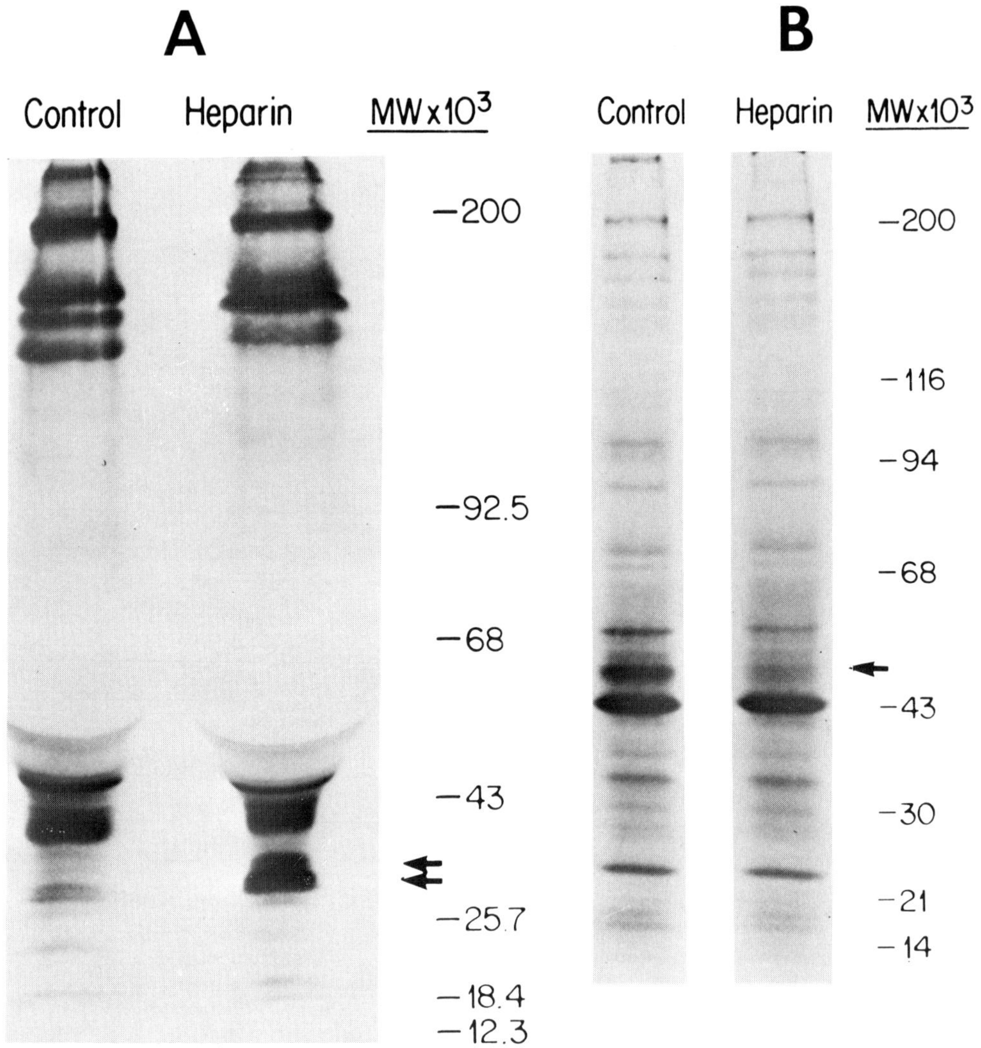

FIGURE 9. Alterations in specific protein synthesis by heparin. Proteins synthesized by serum-stimulated rat aortic SMC 4 hr after release from G_0 in the presence and absence of 100 μg/mℓ heparin were labeled with ^{35}S-methionine and examined by SDS polyacrylamide gel electrophoresis. Note the induction of a doublet in the 35-kdalton region in the culture medium of heparin-treated cells (A, heavy arrows). Also note the disappearance of a 48-kdalton protein in the substrate-attached material of heparin-treated SMC (B, heavy arrow).

uptake of the radiolabeled precursors. Using standard methods for measuring metabolite uptake into cells, we found that heparin inhibited thymidine and uridine uptake by 70 to 80% in SMC.[20] This degree of inhibition is sufficient to account entirely for the observed decrease in DNA and RNA synthesis. Aminoisobutyric acid (a leucine analog) and 3-*O*-methylglucose (a glucose analog) transport was unaffected by heparin. These metabolites were used because they are transported but not further metabolized by cells. In the case of thymidine and uridine, controls were performed to ensure that the radioactivity measured in the uptake assays was not due to incorporation of the metabolites into macromolecules by comparing trichloroacetic acid-soluble and trichloroacetic acid-insoluble counts. Less than 2% of the internalized metabolites were incorporated into macromolecules at the end of the assay period. Although a causal relationship, if any, between the inhibition of nucleoside uptake and the antiproliferative effect of heparin has not been established, this effect of heparin may provide valuable clues to the molecular mechanisms controlling the antiproliferative action of heparin. In addition, these two activities share many characteristics of biological and chemical specificity: (1) the effect is reversible, (2) the effect is specific for SMC, (3) AC and NAC heparin are equally effective, (4) the effect is lost with time in culture, (5) the effect is growth state specific, and (6) heparin is the most potent glycosaminoglycan in producing the effect.

C. Heparin Induces and Represses the Synthesis of Specific Proteins

Although the experiments described previously showed that overall rates of protein synthesis were not affected by heparin, we considered the possibility that heparin could selectively alter the synthesis of specific proteins.[41] SMC from rat aorta, rat mesenteric artery, and Cebus monkey aorta were first growth-arrested and then released from the G_0 block by placing them in normal growth medium with or without heparin. At intervals ranging from 0 to 144 hr after release, the cells were pulsed for 2 hr with ^{35}S-methionine. Incorporation of label into acid-insoluble pools was determined for the following pools: (1) *culture medium* was removed and added to a cocktail containing several protease inhibitors; protein was precipitated with 10% trichloroacetic acid, and the insoluble proteins were dissolved in 4% sodium dodecyl sulfate (SDS) for 5 min at 37°C; (2) *cells* were washed three times with buffer containing protease inhibitors and then removed from the dish by the addition of 1.0 m*M* EGTA with gentle rocking for 30 to 45 min at 37°C; and (3) *substrate-attached material* was prepared by adding 4% SDS to the culture dishes after the cells had been removed; the dishes were incubated with gentle rocking for 1 hr at 37°C and scraped with a rubber policeman.

We detected the induction into the culture medium of a pair of acidic (pI 5.0 to 5.5) proteins of approximately 35 kdaltons in heparin-treated SMC[41] (Figure 9). The induction of the 35-kdalton proteins was inhibited by actinomycin D, and was not affected by treatment with collagenase. Many characteristics of this protein alteration parallel the growth inhibitory property of heparin, including the reversibility, target cell specificity, glycosaminoglycan specificity, effectiveness of NAC heparin, growth state specificity, and loss of effect with time in culture. Furthermore, heparin caused a transient suppression of a 48-kdalton substrate-attached protein. We have not characterized this protein as extensively as the 35-kdalton proteins, but the 48-kdalton protein also appears to be noncollagenous and is most effectively suppressed by heparin compared to other glycosaminoglycans. Recently, Majack and Bornstein[42] have observed the induction of a novel 60-kdalton collagen in the cell fraction of heparin-treated SMC. They have also detected a pair of 37-kdalton proteins in the culture medium of heparin-treated SMC.[43] While it is quite likely that the 35- and 37-kdalton culture medium proteins are the same, to date this has not been directly established. It must be noted that, as is the case for the inhibition of nucleoside uptake, a direct relationship between these specific protein changes and the antiproliferative activity of heparin has not been proven.

The above findings demonstrate that heparin can both induce and suppress the synthesis of specific proteins. The fact that an inhibitor of cell growth can alter the synthesis of selected excreted and intracellular proteins may provide an important communications mechanism for coordinating growth within the vascular wall.

D. Heparin Inhibits Growth of Glomerular Mesangial Cells In Vivo and In Vitro

The glomerular mesangial cell is closely related to the vascular SMC by several criteria, including contractility,[44] ultrastructure,[45] and the presence of SMC-specific myosin.[46] Because this cell proliferates abnormally in certain kidney diseases, we tested the ability of heparin to inhibit the growth of this cell type in vivo and in vitro.

Using Habu snake venom to induce glomerular injury and subsequent mesangial hyperplasia,[47,48] we found that the administration of AC heparin beginning 18 hr after Habu venom resulted in a dramatic decrease in the percentage of injured glomeruli with mesangial cell proliferation. NAC heparin was only partially effective; the reasons for this are not yet known, but could be due to inherent differences in antiproliferative mechanism, inherent differences between mesangial and SMC, or to altered metabolism and clearance of AC and NAC heparins. Olson, using a renal ablation model for kidney injury, also found that heparin protected against such injury,[49] although a direct effect on mesangial cells was not determined.

Cultured mesangial cells were also inhibited by heparin.[50] The antiproliferative effect on these cells was similar to SMC in that the effect was reversible and it was also specific for heparin. Unlike the in vivo situation, cultured mesangial were inhibited by NAC heparin. Interestingly, glomerular epithelial cells were found to secrete a heparin-like inhibitor of mesangial cell growth.[50] In addition, a peptide inhibitor was also excreted by these cells. To make the analogy to the vascular wall more complete, the epithelial cells also secreted a mitogen for mesangial cells. Depending on the precise culture and postcollection treatment conditions, either the inhibitory or the growth-promoting effect can predominate. Much more work is needed to fully understand the precise interrelationships between the growth effectors operating in this system.

VI. A MODEL FOR GROWTH REGULATION IN THE VASCULAR WALL

Based on the results described above, we have postulated that heparin or heparin-like molecules may function as physiological growth regulators in the vascular wall (see also Chapter 4). In the normal, healthy vessel heparin is secreted by EC, SMC, or both into the extracellular matrix, where it functions to help maintain SMC in a quiescent growth state. When the endothelium is damaged, this growth control mechanism is lost, and the medial SMC react to the presence of mitogens by migrating into the intima and proliferating. A key point in this hypothesis is that SMC will not respond to PDGF, macrophage, or other mitogens in the presence of heparin. Indeed, we have recently shown[51] that macrophages from hypercholesterolemic rats strongly stimulate the proliferation of SMC. This effect is abolished by heparin. Therefore, heparin present in the arterial wall must be degraded or otherwise inactivated before SMC can undergo the hyperplastic response. There are at least four possible mechanisms for overcoming the antiproliferative effect of heparin, none of them mutually exclusive: (1) damage to or loss of endothelium may remove one source of heparin; (2) endoglycosidases released from platelets adhering to the injured site or from SMC might cleave cell-bound or matrix-bound heparin into soluble fragments that would be diluted or washed away because endothelial damage has removed a physical barrier to increased hydraulic flux; support for this mechanism comes from studies in our laboratory[52] showing that an endoglycosidase isolated from platelets[53] is capable of cleaving heparin from the EC surface in vitro; (3) the local concentration of mitogens may be so high that SMC can proliferate despite the presence of heparin-like species; and (4) other molecules which have a high affinity for heparin, such as Platelet Factor IV, could inactivate heparin in situ; the observations that Platelet Factor IV penetrates into an injured region in vivo[54] and that this substance blocks the antiproliferative effect of heparin in vitro[72] lend credence to this possibility. When the injured area is re-endothelialized, SMC proliferation appears to cease. This might be due to the elaboration of heparin by EC.

Until now, most of the results discussed were obtained in our laboratory. However, many other investigators have provided experimental support for the above model. For example, Benitz et al.[55] have recently demonstrated that heparin is a potent inhibitor of vascular SMC isolated from fetal pulmonary artery. The characteristics of the inhibition of this SMC type is similar to the inhibition of other types of vascular SMC. These findings may have important implications for the treatment of primary fetal hypertension.

Several laboratories[56-58] have shown that EC secrete heparan sulfates (the class of glycosaminoglycan to which heparin belongs). About half of all the glycosaminoglycans synthesized by EC are heparin sulfates secreted into the culture medium. In addition, the heparan sulfates made by EC were more heparin-like (i.e., more highly sulfated) than the heparan sulfates made by SMC. In contrast to EC, SMC and fibroblasts secrete only about 10% of their complex carbohydrates as heparan sulfates.[59,60]

Other observations implicate heparin-like species and EC as growth regulators in the arterial wall. Eisenstein et al.[61] prepared fractions of bovine aorta extracts and tested their ability to alter the growth of cultured SMC. Both positive and negative growth effects were observed. One of the fractions that inhibited SMC growth contained a heparin-like molecule. It has not been shown that this inhibitory activity was actually mediated by heparin, but in light of our previous results, it is possible that the active material in this fraction of the aortic extract was indeed heparin.

Changes in the glycosaminoglycans of arteries after experimentally induced injury have been reported by several groups,[62-65] and these observations are generally supportive of our hypothesis. Wight et al.[65] carried out careful biochemical analyses of the glycosaminoglycan profile in rabbit aortic intimas after endothelial denudation. Areas that had been reendothelialized accumulated much more glycosaminoglycan than nonendothelialized regions. Most of this accumulation consisted of heparan and chondroitin sulfates. Although the source of the heparan sulfate could not be determined, their results implicate EC in influencing both the amount and class of complex carbohydrate present in the arterial intima.

Ausprunk and her collaborators[66,67] have reported on the ultrastructural localization of glycosaminoglycans in both actively proliferating and quiescent microvascular beds in the rabbit eye. These investigators used ruthenium red to stain complex carbohydrates combined with selective degradation by glycosaminoglycan-degrading enzymes. They found that in actively developing vascular beds the predominant species was hyaluronic acid. In contrast, in quiescent vascular beds the predominant species were heparan sulfates that were found surrounding the SMC, as well as on the abluminal side of the EC. These observations in small vessels are consistent with the putative role of heparin-like molecules in regulating SMC growth in large vessels because the heparan sulfates were found under the quiescent EC and around the SMC; they also suggest that the growth state of EC — proliferative vs. quiescent — plays a crucial role in determining the amount of heparan sulfate present. Conceptually, this agrees with our own data, in which we have shown that confluent, but not proliferating, EC produce a heparin-like inhibitor of SMC.

Another example of the potential relationship between heparin-like compounds and the cessation of growth comes from the work of Bernfield and colleagues on the development of embryonic mouse salivary and mammary tissue.[68,69] In these systems, the predominant glycosaminoglycan in proliferating regions of the tissue is hyaluronic acid, whereas in terminally differentiated regions the major species are heparan sulfates. It is possible that the accumulation of heparin-like species is a control mechanism in development.

VII. POTENTIAL CLINICAL USES OF NAC HEPARIN

The development of NAC, antiproliferative heparin analogs may have clinical significance. A substantial number (perhaps 15%) of vascular surgical procedures such as grafts, shunts, bypasses, angioplasties, and endarterectomies fail within a year, apparently due to SMC proliferation.[70] These surgical failures might be preventable by treating patients with NAC, antiproliferative heparin compounds, thus avoiding hemorrhagic complications. In addition, such heparin derivatives might be useful in the treatment of primary fetal hypertension, the most common cardiovascular defect in newborns, and in the treatment of certain kidney diseases which are characterized by the proliferation of mesangial cells.

REFERENCES

1. Schwartz, S. M., Gajdusek, C. M., and Owens, G. K., Vessel growth control, in *Pathobiology of the Endothelial Cell,* Nossel, H. L. and Vogel, H. T., Eds., Academic Press, New York, 1982, 63.
2. Ross, R., Glomset, J., Kariya, B., and Harker, L., A platelet-dependent serum factor that stimulates the proliferation of arterial smooth muscle cells in vitro, *Proc. Natl. Acad. Sci. U.S.A.,* 71, 1207, 1974.
3. Grotendorst, G. R., Seppa, H. E. J., Kleinman, H. K., and Martin, G. R., Attachment of smooth muscle cells to collagen and their migration toward platelet-derived growth factor, *Proc. Natl. Acad. Sci. U.S.A.,* 78, 3669, 1981.
4. Greenberg, G. B. and Hunt, T. K., The proliferative response in vitro of vascular endothelial and smooth muscle cells exposed to wound fluids and macrophages, *J. Cell. Physiol.,* 97, 353, 1978.
5. Martin, B. M., Gimbrone, M. A., Unanue, F. R., and Cotran, R. S., Stimulation of nonlymphoid mesenchymal cell proliferation by a macrophage-derived growth factor, *J. Immunol.,* 126, 1510, 1981.
6. Gajdusek, C. M., DiCorleto, P., Ross, R., and Schwartz, S. M., An endothelial cell-derived growth factor, *J. Cell Biol.,* 85, 467, 1980.
7. Fishman, J. A., Ryan, G. B., and Karnovsky, M. J., Endothelial regeneration in the rat carotid artery and the significance of endothelial denudation in the pathogenesis of myointimal thickening, *Lab. Invest.,* 32, 339, 1975.
8. Moore, S., Thromboatherosclerosis in normolipemic rabbits. A result of continued endothelial damage, *Lab. Invest.,* 29, 478, 1973.
9. Poole, J. C. F., Cromwell, S. B., and Benditt, E. P., Behavior of smooth muscle cells and formation of extracellular structures in the reaction of arterial walls to injury, *Am. J. Pathol.,* 62, 391, 1970.
10. Ross, R. and Glomset, J. A., Atherosclerosis and the arterial smooth muscle cell, *Science,* 180, 1332, 1973.
11. Spaet, T. H., Stemerman, M. B., Veith, F. J., and Lejnieks, I., Intimal injury and regrowth in the rabbit aorta: medial smooth muscles as a source of neointima, *Circ. Res.,* 36, 58, 1975.
12. Schwartz, S. M., Haudenschild, C. C., and Eddy, E. M., Endothelial regerneration. I. Quantitative analysis of initial stages of endothelial regeneration in rat aortic intima, *Lab. Invest.,* 38, 568, 1978.
13. Chen, L. B. and Buchanan, J. M., Mitogenic activity of blood components. I. Thrombin and prothrombin, *Proc. Natl. Acad. Sci. U.S.A.,* 72, 131, 1975.
14. Clowes, A. W. and Karnovsky, M. J., Suppression by heparin of smooth muscle cell proliferation in injured arteries, *Nature (London),* 265, 625, 1977.
15. Lam, L. H., Silbert, J. E., and Rosenberg, R. D., The separation of active and inactive forms of heparin, *Biochem. Biophys. Res. Commun.,* 69, 570, 1976.
16. Hook, M., Bjork, I., Hopwood, J., and Lindahl, U., Anticoagulant activity of heparin: separation of high activity and low activity species by affinity chromatography on immobilized antithrombin, *FEBS Lett.,* 69, 51, 1976.
17. Guyton, J. R., Rosenberg, R. D., Clowes, A. W., and Karnovsky, M. J., Inhibition of rat arterial smooth muscle cell proliferation by heparin. In vivo studies with anticoagulant and non-anticoagulant heparin, *Circ. Res.,* 46, 625, 1980.
18. Hoover, R. L., Rosenberg, R. D., Haering, W., and Karnovsky, M. J., Inhibition of rat arterial smooth muscle cell proliferation by heparin. II. In vitro studies, *Circ. Res.,* 47, 578, 1980.

19. Castellot, J. J., Addonizio, M. L., Rosenberg, R. D., and Karnovsky, M. J., Cultured endothelial cells produce a heparin-like inhibitor of smooth muscle cell growth, *J. Cell Biol.*, 90, 372, 1981.
20. Castellot, J. J., Cochran, D. L., and Karnovsky, M. J., Effect of heparin on vascular smooth muscle cells. I. Cell metabolism, *J. Cell. Physiol.*, 124, 21, 1985.
21. Lippman, M. and Mathews, M. B., Heparins: varying effects on cell proliferation in vitro and lack of correlation with anticoagulant activity, *Fed. Proc.*, 36, 55, 1977.
22. Fritze, I. M. S., Reilly, C. F., and Rosenberg, R. D., An antiproliferative heparan sulfate species produced by post-confluent smooth muscle cells, *J. Cell Biol.*, 100, 1041, 1985.
23. Oosta, G. M., Gardner, W. T., Beeler, D. L., and Rosenberg, R. D., Multiple functional domains of the heparin molecule, *Proc. Natl. Acad. Sci. U.S.A.*, 78, 829, 1981.
24. Choay, J., Petitou, M., Lormeau, J. C., Sinay, P., Casu, B., and Gatti, G., Structure-function relationships in heparin: a synthetic pentasaccharide with high affinity for antithrombin III and eliciting high anti-factor Xa activity, *Biochem. Biophys. Res. Commun.*, 116, 492, 1983.
25. Kazatchkine, M. D., Fearon, D. T., Metcalfe, D. D., Rosenberg, R. D., and Austen, K. F., Structural determinants of the capacity of heparin to inhibit formation of the human amplification C3 convertase, *J. Clin. Invest.*, 67, 223, 1981.
26. Castellot, J. J., Beeler, D. L., Rosenberg, R. D., and Karnovsky, M. J., Structural determinants of the capacity of heparin to inhibit the proliferation of vascular smooth muscle cells, *J. Cell. Physiol.*, 120, 315, 1984.
27. Castellot, J. J., Wong, K., Herman, B., Hoover, R. L., Albertini, D. F., Wright, T. C., Caleb, B. L., and Karnovsky, M. J., Binding and internalization of heparin by vascular smooth muscle cells, *J. Cell. Physiol.*, 124, 13, 1985.
28. Yamada, K. M., Cell surface interactions with extracellular materials, *Annu. Rev. Biochem.*, 52, 761, 1983.
29. Pastan, I. H. and Willingham, M. C., Journey to the center of the cell: role of the receptosome, *Science*, 214, 504, 1981.
30. Brown, M. S. and Goldstein, J. L., Receptor-mediated endocytosis: insights from the lipoprotein receptor system, *Proc. Natl. Acad. Sci. U.S.A.*, 76, 3330, 1979.
31. Carpenter, G. and Cohen, S., ^{125}I-labeled human epidermal growth factor: binding, internalization and degradation in human fibroblasts, *J. Cell Biol.*, 71, 159, 1976.
32. Karin, M. and Mintz, B., Receptor-mediated endocytosis of transferrin in developmentally totipotent mouse teratocarcinoma stem cells, *J. Biol. Chem.*, 256, 3245, 1981.
33. Helenius, A., Kartenbeck, J., Simons, K., and Fries, E., On the entry of Semliki Forest virus into BHK-21 cells, *J. Cell Biol.*, 84, 404, 1980.
34. Salisbury, J. L., Condeelis, J. S., and Satir, P., Role of coated vesicles, microfilaments, and calmodulin in receptor-mediated endocytosis of cultured B lymphoblastoid cells, *J. Cell Biol.*, 87, 132, 1980.
35. Dunn, W. A. and Hubbard, A. L., Receptor-mediated endocytosis of epidermal growth factor by hepatocytes in the perfused rat liver: ligand and receptor dynamics, *J. Cell Biol.*, 98, 2148, 1984.
36. Robinson, M. S., Rhodes, J. A., and Albertini, D. F., Slow internalization of human chorionic gonadotropin by cultured granulosa cells, *J. Cell. Physiol.*, 117, 43, 1984.
37. Saito, H. and Uzman, B. G., Uptake of chondroitin sulfate by mammalian cells in culture, *Exp. Cell Res.*, 66, 97, 1971.
38. Kresse, H., Tekolf, W., von Figura, K., and Buddecke, E., Metabolism of sulfated glycosaminoglycans in cultivated bovine arterial cells. II. Quantitative studies on the uptake of $^{35}SO_4$-labeled proteoglycans, *Hoppe-Seyler's Z. Physiol. Chem.*, 356, 943, 1975.
39. Wasteson, A., Lindahl, U., and Hallen, A., Mode of degradation of the chondroitin sulfate proteoglycan in rat costal cartilage, *Biochem. J.*, 130, 729, 1972.
40. Pledger, W. J., Stiles, C., Antoniades, H., and Scher, C., Induction of DNA synthesis in Balb/c3T3 cells by serum components: reevaluation of the commitment process, *Proc. Natl. Acad. Sci. U.S.A.*, 74, 4481, 1977.
41. Cochran, D. L., Castellot, J. J., and Karnovsky, M. J., Effect of heparin on vascular smooth muscle cells. II. Specific protein synthesis, *J. Cell. Physiol.*, 124, 29, 1985.
42. Majack, R. A. and Bornstein, P., Heparin-like glycosaminoglycans regulate the phenotype of cultured vascular smooth muscle cells. Induced synthesis of a novel short chain collagen, *J. Cell Biol.*, 97, 2a, 1983.
43. Majack, R. A. and Bornstein, P., Heparin and related glycosaminoglycans modulate the secretory phenotype of vascular smooth muscle cells, *J. Cell Biol.*, 99, 1688, 1984.

44. Ausiello, D. A., Kreisberg, J. I., Roy, C., and Karnovsky, M. J., Contraction of cultured rat glomerular mesangial cells after stimulation with angiotensin II and arginine vasopressin, *J. Clin. Invest.,* 65, 754, 1980.
45. Latta, H. and Maunsbach, A. B., Relations of the centrolobular region of the glomerulus to the juxtaglomerular apparatus, *J. Ultrastruct. Res.,* 6, 563, 1962.
46. Harper, P. A., Robinson, J. M., Hoover, R. L., Wright, T. C., and Karnovsky, M. J., Improved methods for culturing rat glomerular cells, *Kidney Int.,* 26, 91, 1984.
47. Bradfield, J. W. B. and Cattell, V., The mesangial cell in glomerulonephritis. I. Mechanisms of hypercellularity in experimental immune complex disease, *Lab. Invest.,* 36, 481, 1977.
48. Coffey, A. K. and Karnovsky, M. J., Heparin inhibits mesangial cell proliferation in Habu venom induced glomerular injury, *Am. J. Pathol.,* 120, 248, 1985.
49. Olson, J. L., Role of heparin as a protective agent following reduction of renal mass, *Kidney Int.,* 25, 376, 1984.
50. Castellot, J. J., Harper, P. A., Hoover, R. L., and Karnovsky, M. J., Glomerular epithelial cells secrete a heparin-like inhibitor of mesangial cell proliferation, *Am. J. Pathol.,* 120, 427, 1985.
51. Karnovsky, M. J., Castellot, J. J., Robinson, J. M., and Hoover, R. L., Properties of macrophages from hypercholesterolemic rats, *J. Cell Biol.,* 99, 327a, 1984.
52. Castellot, J. J., Favreau, L., Rosenberg, R. D., and Karnovsky, M. J., Inhibition of vascular smooth muscle cell growth by endothelial cell-derived heparin: possible role of platelet endoglycosidase, *J. Biol. Chem.,* 257, 11256, 1982.
53. Oosta, G. M., Favreau, L., and Rosenberg, R. D., Purification and properties of a platelet endoglycosidase, *J. Biol. Chem.,* 257, 11249, 1982.
54. Goldberg, I. D., Stemerman, M. B., and Handin, R. I., Vascular permeation of platelet factor 4 after endothelial injury, *Science,* 209, 611, 1980.
55. Benitz, W. E., Coulson, J. D., Lessler, D. S., Kasnowski, C., and Bernfield, M., Smooth muscle cells from fetal bovine pulmonary arteries in vitro: growth inhibition by heparin, *J. Clin. Invest.,* in press, 1985.
56. Buonassisi, V., Sulfated mucopolysaccharide synthesis and secretion in endothelial cell cultures, *Exp. Cell Res.,* 76, 363, 1973.
57. Busch, C., Ljungman, C., Heldin, C. M., Waskson, E., and Obrink, B., Surface properties of cultured endothelial cells, *Haemostasis,* 8, 142, 1979.
58. Gamse, G., Fromme, H. G., and Kresse, H., Metabolism of sulfated glycosaminoglycans in cultured endothelial cells and smooth muscle cells from bovine aorta, *Biochem. Biophys. Acta,* 544, 514, 1978.
59. Kraemer, P. M., Heparan sulfate of cultured cells. II. Acid-soluble and -precipitable species of different cell lines, *Biochemistry,* 10, 1445, 1971.
60. Wight, T. N. and Ross, R., Proteoglycans in primate arteries. II. Synthesis and secretion of glycosaminoglycans by arterial smooth muscle cells in culture, *J. Cell Biol.,* 67, 675, 1975.
61. Eisenstein, R., Schumacher, B., and Matijevitch, B., Intrinsic regulators of arterial cell growth, *Fed. Proc.,* 38, 1075a, 1979.
62. Minick, C. R., Litrenta, M. M., Alonso, B. R., Silane, M. F., and Stemerman, M. B., Further studies on the effect of regenerated endothelium on intimal lipid accumulation, *Prog. Biochem. Pharmacol.,* 13, 115, 1977.
63. Richardson, M., Ihnatowycz, I., and Moore, S., Glycosaminoglycan distribution in rabbit aortic wall following balloon catheter deendothelialization: an ultrastructural study, *Lab. Invest.,* 43, 509, 1980.
64. Stevens, R. L., Colombo, M., Gonzales, J. J., Hollander, W., and Schmid, K., The glycosaminoglycans of the human artery and their changes in atherosclerosis, *J. Clin. Invest.,* 58, 470, 1976.
65. Wight, T. N., Curwen, K. D., Holman, W., and Minick, C. R., Effect of regenerated endothelium on glycosaminoglycan accumulation in the arterial wall, *Fed. Proc.,* 38, 1075a, 1979.
66. Ausprunk, D. H., Boudreau, C. L., and Nelson, D. A., Proteoglycans in the microvasculature. I. Histochemical localization in microvessels of the rabbit eye, *Am. J. Pathol.,* 103, 353, 1981.
67. Ausprunk, D. H., Boudreau, C. L., and Nelson, D. A., Proteoglycans in the microvasculature. II. Histochemical localization in proliferating capillaries of the rabbit cornea, *Am. J. Pathol.,* 103, 367, 1981.
68. Bernfield, M. R., Banerjee, S. D., and Cohn, R. H., Dependence of salivary epithelial morphology and branching morphogenesis upon acid mucopolysacharide-protein (proteoglycan) at the epithelial cell surface, *J. Cell Biol.,* 52, 674, 1972.
69. David, G. and Bernfield, M. R., Type I collagen reduces the degradation of basal lamina proteoglycan by mammary epithelial cells, *J. Cell Biol.,* 91, 281, 1981.
70. Whittemore, A. D., Clowes, A. W., Couch, N. P., and Mannick, J. A., Secondary femoropopliteal reconstruction, *Ann. Surg.,* 193, 35, 1981.

71. Castellot, J. J., Jr., Choay, J., Lormeau, J.-C., Petitou, M., Sache, E., and Karnovsky, M. J., Structural determinants of the capacity of heparin to inhibit the proliferation of vascular smooth muscle cells. II. Evidence for a pentasaccaride sequence that contains a 3-0-sulfate, *J. Cell Biol.*, 102, 1979, 1986.
72. Castellot, J. J., Jr. and Karnovsky, M. J., unpublished observation.

Chapter 8

BIOSYNTHESIS AND MODULATION OF EXTRACELLULAR MATRIX COMPONENTS BY CULTURED VASCULAR SMOOTH MUSCLE CELLS

Richard A. Majack and Paul Bornstein

TABLE OF CONTENTS

I. INTRODUCTION

Vascular smooth muscle cells (SMC) are the predominant cell type in the media of the blood vessel wall and serve both a major contractile function and an important synthetic function. SMC produce a spectrum of products which regulate vascular tone and platelet aggregation.[1-7] In addition, SMC produce a variety of molecules which, in the aggregate, comprise the extracellular matrix (ECM). This highly developed ECM is composed of proteoglycans, glycoproteins (such as thrombospondin and fibronectin), elastin, and at least five genetically distinct types of collagen. SMC also produce enzymes capable of degrading elastin[8] and presumably other matrix components. SMC-derived proteases (such as collagenases and elastases) undoubtedly play important roles in cellular movement, tissue remodeling and development, and in the regulation of production of ECM molecules.

In this review we focus on recent in vitro studies of the production and modulation of ECM components by vascular SMC and the reciprocal modulation of SMC functions by the ECM. The SMC connective tissue matrix, in vivo, forms the structural framework of the vessel wall and determines, in part, its mechanical properties. Matrix components can act as a thrombogenic surface,[9] can sequester biologically active molecules such as growth factors,[10] and may serve as important determinants of SMC function (e.g., growth, migration, and differentiation). The production of matrix molecules by vascular SMC is undoubtedly altered both quantitatively and qualitatively during hypertension, atherosclerosis, angiogenesis, and early development. Modulation of ECM components in vitro, as discussed here, presumably reflects similar mechanisms which operate in vivo and which may alter the structural and/or functional properties of the vessel wall and the cells therein.

II. MATRIX COMPONENTS SYNTHESIZED BY CULTURED VASCULAR SMOOTH MUSCLE CELLS

Vascular SMC from a variety of species produce an extensive ECM in culture.[11] The primary components of the SMC matrix include fibronectin, thrombospondin, elastin, and collagen; these macromolecules will be discussed below. The production and modulation of proteoglycans by vascular SMC will not be covered in this chapter.

A. Fibronectin

Fibronectins are high molecular weight, dimeric modular glycoproteins (subunit M_r 200,000 to 250,000) found in body fluids and in connective tissue matrices. The structure, distribution, and biological functions of the fibronectins have been the subject of several recent reviews.[12-15] Fibronectins contain structurally distinct domains which interact with collagen, fibrin, cell surfaces, and heparan sulfate[12-14] and play important roles in cell adhesion, spreading, proliferation, cytoskeletal organization, migration, chemoattraction, phagocytosis, and differentiation.[12,13,15] The synthesis of fibronectin by SMC in vitro has been established by metabolic labeling of monkey[16] and porcine[17] aortic SMC and by detection with antibodies in cultures of monkey aortic SMC.[18] We have determined, using immunoisolation techniques, that rat aortic SMC also produce fibronectin in culture (see Figures 1 and 2). The function of fibronectin in SMC cultures may encompass the wide range of functions ascribed to the fibronectins in general. Fibronectin is known to act as an attachment factor for guinea pig and monkey SMC in vitro.[19] In addition, fetal sheep SMC use fibronectin to attach to collagen types I and III.[20] Fibronectin may also play a role in the control of "nodulation" in porcine aortic SMC cultures.[17,21]

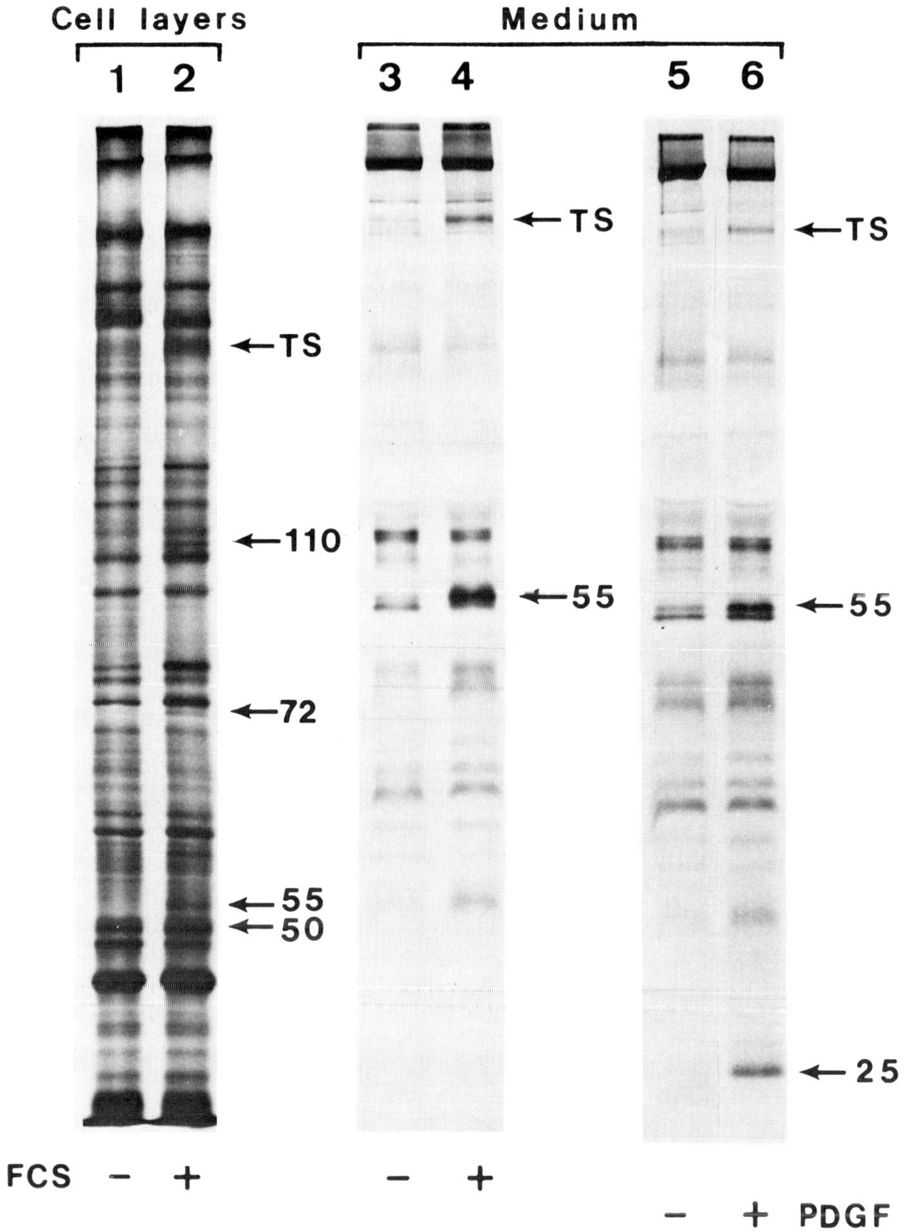

FIGURE 1. Modulation of SMC biosynthetic phenotype by serum and PDGF. Rat aortic SMC were maintained in a growth factor-deficient state by culture in 10% plasma-derived serum (PDS). PDS-SMC were metabolically labeled with [^{35}S]methionine in methionine-free Waymouth's medium for 1 hr at 37°C or following a 1 hr treatment with 10% fetal calf serum (FCS) or 5 ng/mℓ platelet-derived growth factor (PDGF). Cell layer and medium proteins were harvested into protease inhibitors and resolved by SDS-PAGE under reducing conditions using 8 (lanes 1 and 2) or 10% (lanes 3 to 6) separating gels. Proteins were visualized by autoradiography. In the cell layer, FCS consistently stimulated the production of thrombospondin (TS) and protein chains with approximate M_r 50,000, 55,000, 72,000, and 110,000 (lanes 1 and 2). TS and the M_r 55,000 protein were secreted into the culture medium (lanes 3 and 4). The effect of FCS on SMC secreted proteins could be completely mimicked by treatment of PDS-SMC with highly purified PDGF (kindly provided by E. Raines and Dr. R. Ross). Exposure of PDS-SMC to PDGF also induced the production of an M_r 25,000 secreted protein not observed in FCS-stimulated SMC (lane 6). Molecular weights (10^{-3}) of the induced proteins are indicated at arrows.

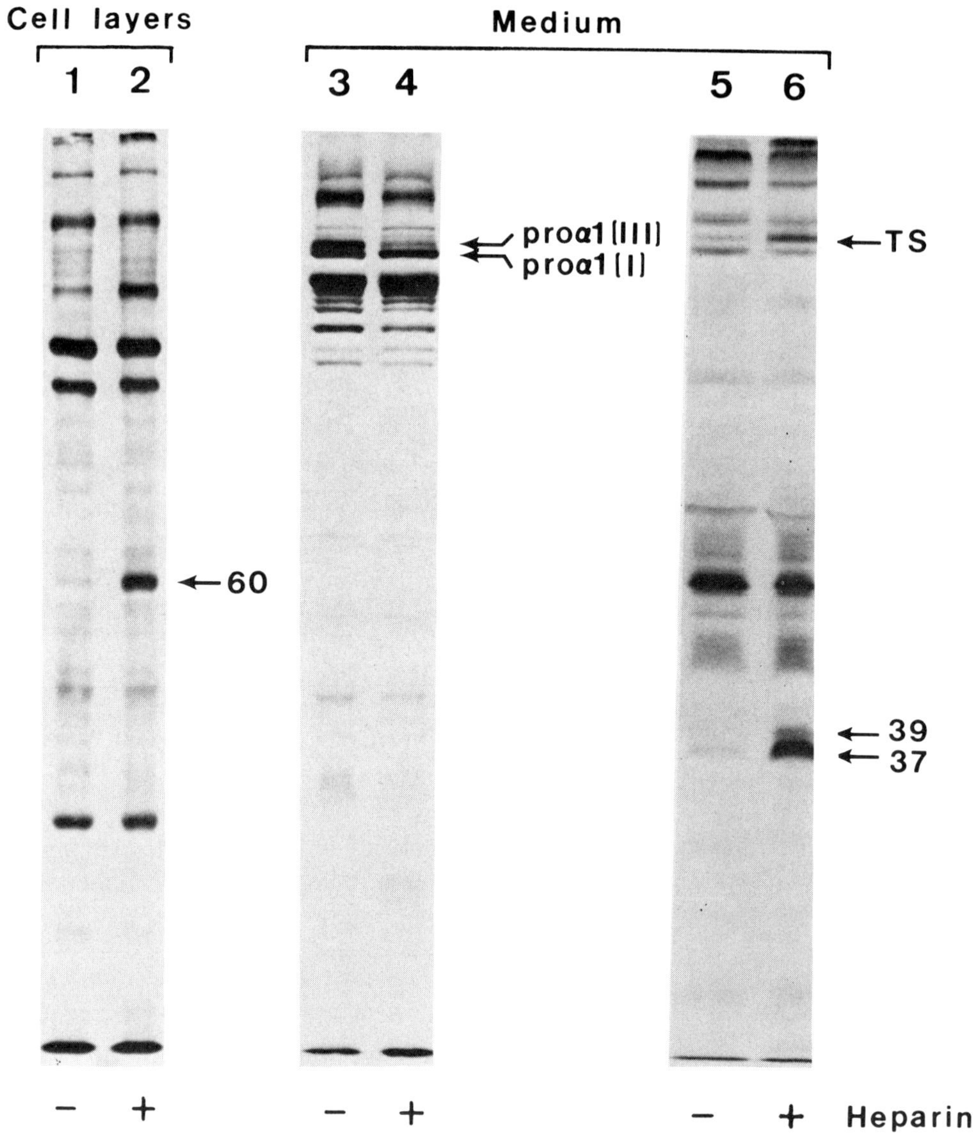

FIGURE 2. Modulation of SMC biosynthetic phenotype by heparin. Rat aortic SMC were grown to over-confluence and maintained in medium containing 10% FSC. Following exposure of SMC to 100 μg/mℓ heparin, cell proteins were metabolically labeled with [^{3}H]proline or [^{35}S]methionine and visualized by SDS-PAGE on 8 (lanes 1 to 4) or 10% (lanes 5 and 6) separating gels as described in the legend for Figure 1. SMC were exposed to heparin for 72 hr, then labeled with [^{3}H]proline for 24 hr in the continued presence of heparin (lanes 1 to 4). Heparin-treated SMC synthesized increased amounts of a cell-layer-associated short chain (M_r 60,000) collagenous protein and reduced the production of secreted type III procollagen (proα1(III)). In other experiments, SMC were exposed to heparin during a short (2 hr) labeling period with [^{35}S]methionine (lanes 5 and 6). In the presence of heparin, SMC synthesized and secreted increased amounts of two noncollagenous proteins (M_r 37,000 and 39,000) and augmented the release of thrombospondin (TS) into the culture medium. Details are given in the text. Molecular weights ($\times 10^{-3}$) are given for the proteins indicated at arrows; the position of migration of the proα1 chain of type I procollagen (proα1(I)) is also shown.

B. Thrombospondin

Thrombospondin (TS) is an M_r 450,000 glycoprotein first described as a component of platelet α granules[22,23] but now known to be synthesized by fibroblasts, bovine aortic SMC, and endothelial cells in culture.[24-27] The TS molecule consists of three probably identical disulfide-bonded chains of M_r 145,000,[23,28,29] with a heparin-binding region located in an NH_2-terminal domain.[30-32] Incorporation of TS into the ECM of cultured fibroblasts,[26,27] endothelial cells,[26] and SMC[26] has been demonstrated immunologically. The incorporation of TS into extracellular matrices undoubtedly reflects its ability to interact with heparan sulfate proteoglycan and other matrix components.[33-35] While platelet-derived TS is thought to act as an endogenous platelet lectin,[36] mediating platelet:platelet[36,37] or platelet:connective tissue[34] interactions, the function of cell-derived TS and its role in the ECM has not been elucidated.

TS production by cultured bovine SMC has been documented by metabolic labeling[24,26] and by indirect immunofluorescence.[26] The production of TS by cultured human SMC was found to be dependent on cell density; SMC at sparse density secreted more TS into the culture medium than did confluent cultures.[38] We have recently shown that TS is produced in vitro by rat aortic SMC and that its production is regulated by platelet-derived growth factor (see later).[39] While the function of matrix TS is not known for any cell type, our data suggest that TS participates in some manner in the SMC growth response.

C. Elastin

Elastin is a major structural component of the vessel wall and is synthesized by cultured arterial SMC as an M_r 72,000 to 76,000 protein.[40-42] The newly synthesized elastin (tropoelastin) is soluble but can be converted into insoluble elastin by the formation of unique lysine-derived desmosine and isodesmosine crosslinks.[43] Synthesis of both soluble and insoluble forms of elastin has been demonstrated in cultures of arterial SMC derived from rabbit,[44-47] pig,[48] and monkey.[49] Another component of elastic fibers, the microfibrillar protein,[50] has been identified morphologically in cultures of monkey, rabbit, pig, guinea pig, and rat SMC.[42,50-53] Biochemical characterization of the microfibrillar protein is incomplete[54] and will not be discussed here. Morphological data suggest that the fibrillar component of elastic fibers may appear first in the ECM and may act as a framework upon which insoluble elastin is deposited.[42,50,51]

D. Collagens

Vascular SMC are now known to produce at least five distinct types of collagen in culture, including types I, III, V, and VI, CP45 (a pepsin-resistant fragment of unknown origin), an M_r-60,000-short-chain collagen, and possibly type IV.[40,55] Collagen production by cultured SMC has been studied quantitatively, by assessing the relative amount of total collagen produced as a fraction of total protein synthesis and qualitatively, by determining the relative amounts of specific collagen types, primarily by SDS-PAGE. The production of collagens by cultured SMC has been shown to be selectively and quantitatively increased (as a percent of total protein synthesis) by cyclic stretching[56] and heparin treatment[57,58] and decreased by treatment of SMC with 17β-estradiol.[59] Cells cultured from atherosclerotic aortas[60] or treated with growth factors[40,61] also produced more collagen per cell, but collagen production relative to total protein synthesis was unchanged.

1. Type I Collagen

The type I collagen molecule consists of three chains with subunit composition

$[\alpha 1(I)]_2\alpha 2(I)$. Each α chain has a molecular weight of 95,000 (see Bornstein and Sage[62] for a recent review of the biosynthesis, processing, and structures of collagen types I to V). Production of type I collagen by SMC in vitro has been extensively documented (see References 40 and 63 to 66 and references therein). Type I is an interstitial collagen which is believed to be essential for normal development of the vessel wall[67] and can serve in culture as a substrate for SMC adhesion.[19,20] The production of type I collagen by bovine SMC was found to be decreased in confluent cultures.[63] This decrease occurred to a greater extent than a concomitant decrease in type III collagen production, thus leading to an increase in the ratio of types III:I in confluent SMC cultures. In contrast, Beldekas et al.[68] described an increase in type I production by bovine SMC with increasing cell density.

2. Type III Collagen

Type III collagen is a trimer with chain composition $[\alpha 1(III)]_3$. The α chains also have individual molecular weights of 95,000 and are products of a single gene. The pathway of processing of type III collagen in cultured bovine SMC appears to be the same as that described for type I; however, the rate of processing of type III procollagen is markedly slower.[69] The production of type III collagen by SMC and a discussion of the inter- and intraspecies discrepancies in the type I:III ratio reported by various workers have been reviewed previously;[40] additional reports of type III synthesis by cultured SMC have subsequently appeared.[59,63-66,68]

The function of type III collagen appears to be related to its distribution as an interstitial collagen. Patients with type IV Ehlers-Danlos syndrome, in which there is a defect in type III collagen synthesis or secretion, show a number of tissue defects, including a tendency to rupture large blood vessels.[70] Production of type III collagen was found to be reduced in heparin-treated SMC,[71] in bovine SMC treated with 17β-estradiol,[59] in confluent as compared with sparse cultures of rabbit SMC,[63] and in guinea pig and monkey aortic SMC after repeated passage in culture.[72] Beldekas et al.,[68] in studies of cultured bovine SMC, have reported an increased production of type III (and type I) procollagen with increasing cell number. Additionally, the relative amount of type III collagen (compared with type I) was increased twofold in dense cultures.

A number of reports have suggested that the relative content of types I and III collagens differs in atherosclerotic plaques, compared with normal aortic media.[73,74] While several reports have indicated a greater proportion of type III collagen in vivo in diseased or proliferative tissues,[73] other studies have indicated the opposite[74] or no change in the type I:III ratio.[75] Attempts to study this question by culturing aortic SMC from control or proliferative tissues have not clarified the issue. Pietila and Nikkari[60] analyzed collagen production by rabbit SMC cultured from control and atherosclerotic aortas and found no specific alterations in collagen production in cells from diseased tissues (see Chapter 13).

3. Type IV Collagen

Type IV collagen is synthesized as two genetically distinct chains of M_r 185,000 and 170,000 and occurs as a heterotrimer of chain composition $[\alpha 1(IV)]_2\alpha 2(IV)$.[76] Type IV is a normal constituent of most basal laminae;[77] its synthesis by SMC is therefore expected in view of the extensive external lamina elaborated by these cells in vivo and in vitro.[11] Sankey and Barnes[78] have presented preliminary evidence that cultured porcine aortic SMC may produce type IV collagen. In their study, a pepsin-resistant, collagenase-sensitive polypeptide (reduced M_r 140,000) was thought to be derived from type IV collagen on the basis of its molecular weight and solubility characteristics.

4. Type V Collagen

The biosynthesis of type V procollagen chains and processing of the molecules to functional extracellular forms are complex processes whose details remain to be established.[79] Several forms of the protein exist including a heterotrimer of three different chains $\alpha 1(V)$, $\alpha 2(V)$, and $\alpha 3(V)$.[80] Mayne et al.[81] demonstrated the presence of the $\alpha 1$ and $\alpha 2$ chains of type V collagen following pepsin digestion of metabolically labeled cultures of rhesus monkey aortic SMC. Type V collagen has also been identified by immunoprecipitation techniques in the medium of cultured rat SMC.[134]

Grotendorst et al.[20] have shown that sheep aortic SMC attach to type V collagen matrices in a fibronectin-independent manner, in contrast to the attachment of SMC to types I and III collagens.[19,20] SMC preferentially attached to matrices composed of the $\alpha 2(V)$ chain (compared to $\alpha 1(V)$), suggesting the presence of an SMC binding site on the $\alpha 2(V)$ collagen chain. The SMC type V "receptor" was found to be trypsin-resistant, neuraminidase-sensitive, and blocked by treatment with concanavalin A. These data, together with immunofluorescence studies indicating a pericellular distribution of type V around SMC *in situ*,[82] suggest an important role for type V collagen in the interaction of SMC with their extracellular environment.

5. Type VI Collagen

Type VI collagen has been characterized primarily as a pepsin-resistant fragment containing three distinct chains,[83-85] but the primary biosynthetic product and tissue forms are substantially larger.[86-88] The production of type VI collagen by cultured human SMC has been detected, using both immunoblotting and immunofluorescence techniques.[89] The presence of type VI collagen in the aortic media was also demonstrated immunologically,[89] suggesting the synthesis of type VI by SMC in vivo.

6. CP45 and the M_r 60,000 SMC Collagen

Mayne et al.[72,81] and Majack and Bornstein[57,58] have reported the production of other collagenous proteins by cultured SMC. Mayne et al.[72,81] prepared pepsin extracts of guinea pig and monkey aortic SMC and found a short-chain (M_r 45,000) pepsin-resistant collagen, which they termed CP45. Production of CP45 was found to be greatest in early-passage SMC. The molecular weight of CP45 was not affected by reducing agents, suggesting an absence of inter- or intramolecular disulfide bonds. Cyanogen bromide peptides of CP45 were not comparable to those of collagen types I, III, or V. Little else is known concerning the size, structure, or nature of the precursor to CP45. We have recently reported that rat SMC produce an M_r 60,000 collagenous protein which is cell-layer associated, heparin-inducible, and digested by pepsin to an M_r 45,000 to 50,000 fragment.[57,58] The relationship of the M_r 60,000 heparin-induced SMC collagen to other known collagen types is not known.

In recent years the number of recognized collagen types has increased dramatically; at least 18 separate gene products have now been described. Among these new collagens are several "short-chain" collagens that are considerably smaller than the well-known interstitial collagens such as types I and III. The M_r 60,000 SMC collagen, based on its molecular weight, may be related to the pepsin resistant collagenous protein referred to as CP45. The M_r 60,000 SMC collagen may also be related to an M_r 59,000 chondrocyte collagen (type X collagen) extensively described from in vitro sources.[90,91] The M_r 60,000 SMC protein is similar to type X collagen in several respects: (1) both have M_r of approximately 60,000, (2) both contain pepsin-resistant fragments of M_r approximately 45,000,[58,90] (3) neither contains disulfide bonds,[90] (4) both are inducible by matrix components,[91] and (5) both contain similar amounts of hydroxyproline.[91] Schmid and Linsenmayer, in a recent abstract,[92] have described the distribution of type X collagen using monoclonal antibodies and immunofluorescence.

Type X collagen was localized to hypertrophied chondrocytes in developing bone, but not to tissues which remain permanently cartilagenous. The collagen was also present in endochondral and membrane bone, but not in any other tissue, suggesting it may not be an SMC protein. The exact relationship, if any, between the M_r 60,000 SMC collagen and type X is therefore unclear and must await future studies.

III. SECRETORY PHENOTYPES OF VASCULAR SMC IN CULTURE: MODULATION BY GROWTH FACTORS AND HEPARIN-LIKE GLYCOSAMINOGLYCANS

SMC migration and proliferation are believed to be essential steps in the early vascular response to injury and consequently may play an important role in atherogenesis. The platelet-derived growth factor (PDGF) is mitogenic,[93] chemokinetic,[94,95] and chemotactic[20,96] for vascular SMC and is now recognized as a major determinant of SMC function, at least in vitro. Recent studies have indicated that other regulatory molecules, such as heparin-like glycosaminoglycans (GAG), may also control certain SMC functions. In contrast to PDGF, these molecules may be continually present in the normal vascular wall[97,98] and are inhibitory to SMC growth and movement. It is likely that PDGF and regulatory GAG play important roles in angiogenesis, normal development, and tissue repair as well.

We are interested in defining the molecular basis for SMC growth regulation. Accordingly, we have analyzed the biosynthetic phenotype of growth factor-deprived, growth factor-stimulated, and heparin-treated rat aortic SMC in culture.

A. Modulation of SMC Phenotype by Factors in Serum

PDGF has been identified as the principal mitogen responsible for SMC growth in vitro.[99] Addition of nanogram quantities of PDGF to fibroblastic cells in culture elicits a pleiotropic response, resulting in many cellular changes.[100,101] These alterations include an overall increase in protein synthesis,[102,103] stimulation of polysome formation,[102] induction of specific mRNAs,[104,105] and the induction of synthesis of specific "competence" proteins.[106] Data on growth factor-induced protein synthesis have been derived, almost exclusively, from studies on fibroblasts. In 3T3 cells, PDGF induces the rapid transcription of several known oncogenes including c-fos[107] and c-myc.[108] PDGF also stimulates the rapid (within 1.5 hr) production of cytoplasmic proteins, detectable by SDS-PAGE after metabolic labeling, with M_r 29,000, 35,000, 45,000, 60,000, and 70,000.[106] Growth factor-inducible secreted proteins of M_r 12,000, 24,000, 29,000, 34,000, 48,000, and 62,000 have also been described from 3T3 cells.[109,110]

We have recently defined the effects of serum and purified growth factors on the synthesis of specific proteins by vascular SMC.[39] The results, as described in part below, suggest that specific cell types may differ qualitatively in their response to growth stimuli. The data also support previous suggestions[111,112] that the character of the ECM may be an important determinant of cellular growth.

Rat aortic SMC were maintained in growth factor-deficient medium by culturing the cells for 4 to 7 days in Waymouth's medium supplemented with 10% plasma-derived fetal bovine serum (PDS), for 4 days in Waymouth's plus 0.5% fetal bovine serum (FCS), or for 4 days in a defined medium as described by Libby and O'Brien.[113] Cells were stimulated for 1 hr with 10% FCS in Waymouth's medium, then metabolically labeled for 1 hr with [^{35}S]methionine (see Reference 39 for protocols). Control cultures were not stimulated with FCS prior to labeling. Cell layer and medium proteins were resolved by SDS-PAGE under reducing conditions and visualized by autoradiography. As shown in Figure 1, the pattern of proteins synthesized by growth-arrested SMC differed qualitatively from that of serum-stimulated cells; the results were qualitatively similar regardless of the method of growth arrest used. In the cell layer (Figure 1, lanes 1 and 2), serum specifically and reproducibly induced the production of proteins with approximate M_r 50,000, 55,000, 72,000, 110,000, and 150,000. Other cell layer proteins were inconsistently increased or decreased following serum stimulation. Serum also altered the pattern of SMC secreted proteins (Figure 1, lanes 3 and 4). Cultures stimulated with FCS secreted increased amounts of proteins with reduced M_r 55,000 and 150,000.

We have identified the protein present in the cell layer and medium compartments with a chain M_r 150,000 as cellular thrombospondin.[39] The M_r 150,000 chain comigrated on SDS-PAGE with human platelet thrombospondin and was immunoprecipitated by antibodies prepared against human platelet or bovine endothelial cell thrombospondin. The induction of synthesis of thrombospondin and the M_r 55,000 protein can be elicited by treating SMC with purified PDGF (Figure 1, lanes 5 and 6) but not with other growth factors (fibroblast growth factor, epidermal growth factor, multiplication stimulating activity, or insulin). PDGF also stimulated the synthesis of an M_r 25,000 secreted protein not detected in the medium of serum-stimulated cells. It should be noted that these experiments were designed to study only the earliest events following growth factor stimulation and that our gel system was optimized for the resolution of higher molecular weight abundant proteins. It is likely that other specific protein synthetic events occur transiently and/or at later times following stimulation of SMC.

Our data suggest that PDGF treatment of vascular SMC may result in the induction of a unique set of "competence" proteins (compare with the PDGF-regulated 3T3 proteins listed above). While it is likely that many factor-regulated proteins may be common to all cells, other inducible proteins (especially matrix components) may be specific for each cell type. The most interesting results of our studies, in terms of the biology of connective tissue molecules, is the observation that PDGF selectively stimulated SMC thrombospondin production. To date the function of cell-derived or matrix-bound thrombospondin has been unclear. Our data suggest that thrombospondin may play some as-yet-unidentified role in the SMC growth response.

The role of ECM in the growth of cells has been the subject of extensive investigation in recent years.[111,112,114,115] It is clear that cells require an appropriate ECM for optimal growth,[116] and that growth factors may regulate the production of specific matrix molecules. Epidermal growth factor and fibroblast growth factor both appear to selectively inhibit the synthesis of type I collagen in several cell types.[117-119] Fibronectin, another component of the ECM of most cells, has been shown to act as a competence factor for HFL-1 fibroblasts.[115] These studies emphasize the hypothesis[114] that one function of growth factors may be to direct the synthesis of a growth-supportive ECM. For vascular SMC, the ECM optimal for cell proliferation and movement may contain thrombospondin.

B. Modulation of SMC Phenotype by Heparin-Like Glycosaminoglycans

In 1977 Clowes and Karnovsky[120] demonstrated that heparin treatment of animals significantly reduced the proliferative response of neointimal SMC following experimental vascular injury in vivo. A follow-up study[121] established that the growth inhibitory effect of heparin was not related to the anticoagulant properties of heparin or to an inhibitory effect of heparin on platelet adherence or endothelial cell regeneration. Subsequent studies, using cultured rat SMC, verified that heparin was a cell-specific inhibitor of SMC proliferation.[122-124]

The importance of these observations was emphasized by data suggesting that heparin-like molecules may be normally present within the vascular wall. Gamse et al.[125] studied GAG synthesized by cultured vascular endothelial and SMC and found that endothelial heparan sulfate molecules were very similar to heparin in charge density and N-sulfate content. Conditioned medium from cultured quiescent aortic endothelium was found to be growth-inhibitory for vascular SMC.[126-128] The inhibitory activity required low concentrations of serum for release and was destroyed specifically by heparinase.[126] The enzyme present in serum that was responsible for release of the inhibitor has been shown to be a platelet-derived heparitinase.[129] This enzyme, which releases heparin-like GAG fragments from proteoglycans, is believed to be present in endothelium and in smooth muscle as well.[130]

SMC also produce heparan sulfate proteoglycans.[125,131] SMC heparan sulfate may also be growth inhibitory for SMC in culture, especially when derived from quiescent confluent SMC cultures rather than from cells in log-phase growth.[130] The above data collectively indicate that (1) vascular wall cells produce heparan sulfate proteoglycans, (2) a heparitinase may be present in normal and injured vessel walls which is capable of cleaving proteoheparan sulfates and generating heparin-like GAG fragments, and (3) these GAG are growth inhibitory to SMC. Thus SMC growth may be controlled, in an unknown fashion, by heparin-like GAG derived from endothelium or SMC (see Chapters 4 and 7).

Recently, Majack and Clowes[132] reported that heparin-like molecules inhibit the migration of rat vascular SMC in vitro, suggesting that other SMC functions may also be regulated by GAG. The mechanism(s) underlying this phenomenon and its potential role in SMC growth regulation remain unclear.

We have recently demonstrated that heparin and related molecules dramatically alter the synthetic phenotype of cultured rat vascular SMC[57,58,71] (Figure 2). Addition of heparin to the culture medium of early passage rat SMC resulted in a marked (3- to 15-fold) increase of a cell-layer associated M_r 60,000 protein which was sensitive to digestion by purified bacterial collagenase and contained significant amounts of hydroxyproline (Figure 2, lanes 1 and 2). Pulse-chase analysis of heparin-treated SMC revealed that the M_r 60,000 collagen was a primary and abundant product of these cells and was not processed extracellularly to a smaller form. The inductive effect of heparin could be mimicked by iota carrageenan or dextran sulfates but not by hyaluronic acid, dermatan sulfate, or chondroitin sulfates. The induction was concentration-dependent with a maximal effect observed at a heparin concentration of 10 μg/mℓ. Synthesis of the M_r 60,000 collagen increased 18 to 24 hr following addition of heparin to the cultures. Following induction and subsequent removal of heparin, synthesis of the protein remained maximal for at least 12 hr and required 72 hr to return to a basal level.

Heparin-treated SMC also produce decreased amounts of type III procollagen relative to type I procollagen[71] (Figure 2, lanes 3 and 4), and synthesize increased amounts of two noncollagenous proteins (M_r 37,000 and 39,000) (Figure 2, lanes 5 and 6). This effect could be mimicked by heparan and dermatan sulfates and iota carrageenan but not by hyaluronic acid, chondroitin-4-sulfate, or chondroitin-6-sulfate. The inductive effect of heparin was dose-dependent and occurred rapidly (within 1 hr) with maximal induction (three- to fivefold over controls) occurring after 10 to 12 hr of treatment. The effect was rapidly reversible (within 1 hr) and was not altered in the presence of actinomycin D, suggesting regulation at a posttranscriptional level. The relationship between the M_r 37,000 and 39,000 heparin-inducible proteins and an M_r 38,000 heparin-binding protein secreted from "nodular" cultures of bovine SMC[133] is not presently known.

We have recently characterized one additional change in the secretory phenotype of heparin-treated SMC.[39] Following heparin treatment, rat SMC released increased amounts of an M_r 150,000 (reduced) protein which we have identified (using specific immunoprecipitation) as cellular thrombospondin. The increased amount of thrombospondin secreted into the culture medium was directly correlated with a concomitant decrease in cell layer-associated thrombospondin. Analysis of Triton-X-100-insoluble

SMC matrices prepared from control and heparin-treated cells revealed that this effect resulted from the inhibition of incorporation of newly synthesized thrombospondin into the SMC extracellular matrix. The release of thrombospondin from cell layers was concentration-dependent, with a half-maximal effect observed at 0.1 μg/mℓ heparin and a maximal effect at heparin concentrations greater than 1.0 μg/mℓ. Dermatan sulfate, several heparan sulfate preparations, and a variety of heparins with a range of anticoagulant activities were effective in altering the distribution of thrombospondin in culture; other GAG tested were not active. Time-course studies indicated that maximal release of newly synthesized thrombospondin from SMC was achieved with 4 hr of heparin addition; the effect was completely reversible within 8 hr of removal of heparin from the cultures. These data suggest that heparin-like molecules may alter the composition and/or structure of the SMC extracellular matrix without direct cellular (i.e., biosynthetic) involvement. The data are especially significant in view of our recent observation (discussed above) that thrombospondin may be involved in the SMC growth response. We have proposed[39] that heparin-mediated inhibition of SMC:thrombospondin interactions may be involved in the inhibition of SMC migration and growth by heparin-like molecules.

IV. CONCLUSION

Vascular SMC maintained in culture produce many known components of the extracellular matrix. Expression of specific ECM molecules in vitro undoubtedly reflects expression of these matrix components at some point during SMC development. Similarly, modulation of production of ECM components in culture may reflect similar processes in vivo and may provide valuable insight into the functional role of these molecules.

In this paper we have described the primary collagen and glycoprotein matrix products of vascular SMC in culture and have noted, where applicable, the conditions under which synthesis of these molecules is modulated. Production of interstitial collagens (types I and III) can be altered by estrogens,[59] cell density,[63] or glycosaminoglycan treatment.[58,71] Synthesis of SMC thrombospondin varies according to cell density,[38] a finding which may be explained by our recent observation that SMC thrombospondin production is regulated by PDGF.[20] Heparin-like molecules induce the production of a novel SMC matrix protein, an M_r 60,000 short-chain collagen.[57,58] In addition, PDGF and heparin-like glycosaminoglycans, which regulate the growth and migration of vascular SMC in an antagonistic fashion, induce the production of a number of uncharacterized SMC proteins which may or may not be associated with the extracellular matrix.[39,71] Finally, heparin-like growth inhibitors alter the secretory phenotype of cultured SMC by regulating the interaction between thrombospondin and the SMC surface or matrix.[39] These observations, taken together, suggest an important role for thrombospondin and other matrix molecules in the control of SMC growth, migration, and differentiation. Future studies of the regulation, structure, molecular interactions, and function of each of these molecules should advance our understanding of the processes underlying atherogenesis, angiogenesis, and the normal development of the vessel wall.

ACKNOWLEDGMENT

Original contributions from this laboratory were supported in part by NIH Grants AM 11248, HL 18645, and DE 02600 and by a grant from R. J. Reynolds, Inc. R. A. Majack was the recipient of a postdoctoral fellowship from training grant NIH HL 67312.

REFERENCES

1. Dejana, E., Balconi, G., de Castellarnau, C., Barbieri, B., Veragara-Dauden, M., and de Gaetano, G., Prostacyclin production by human endothelial and bovine smooth muscle cells in culture, *Biochim. Biophys. Acta,* 750, 261, 1983.
2. Booyse, F. M., Scheinbuks, J., Radek, J., Osikowicz, G., Feder, S., and Quarfoot, A. J., Immunological identification and comparison of plasminogen activator forms in cultured normal human endothelial cells and smooth muscle cells, *Thromb. Res.,* 24, 495, 1981.
3. Tansik, R. L., Namm, D. H., and White, H. L., Synthesis of prostaglandin 6-keto $F_{1\alpha}$ by cultured aortic smooth muscle cells and stimulation of its formation in a coupled system with platelet lysates, *Prostaglandins,* 15, 399, 1978.
4. Baenziger, N. L., Dillender, M. J., and Majerus, P. W., Cultured human skin fibroblasts and arterial cells produce a labile platelet inhibitory prostaglandin, *Biochem. Biophys. Res. Commun.,* 78, 294, 1977.
5. Larrue, J., Rigaud, M., Daret, D., Demond, J., Durand, and Bricand, H., Prostaglandin production by cultured smooth muscle cells from atherosclerotic rabbit aorta, *Nature (London),* 285, 480, 1980.
6. Re, R., Fallon, J. T., Dzau, J., Quay, S. C., and Haber, E., Renin synthesis by canine aortic smooth muscle cells in culture, *Life Sci.,* 30, 99, 1982.
7. Ager, A., Gordon, J. L., Moncada, S., Pearson, J. D., Salmon, J. A., and Trevethick, M. A., Effects of isolation and culture on prostaglandin synthesis by porcine aortic endothelial and smooth muscle cells, *J. Cell. Physiol.,* 110, 9, 1982.
8. Hornebeck, W., Brechemier, D., Bourdillon, M. C., and Robert, L., Isolation and partial characterization of an elastase-like protease from rat aorta smooth muscle cells. Possible role in the regulation of elastin biosynthesis, *Connect. Tissue Res.,* 8, 245, 1981.
9. Sakariassen, K. S., Bolhuis, P. A., and Sixma, J. J., Human platelet adhesion to artery subendothelium is mediated by Factor VIII:VWF bound to subendothelium, *Nature (London),* 279, 636, 1979.
10. Smith, J. C., Singh, J. P., Lillquist, J. S., Goon, D. S., and Stiles, C. D., Growth factors adherent to cell substrate are mitogenically active *in situ, Nature (London),* 296, 154, 1982.
11. Oakes, B. W., Batty, A. C., Handley, C. J., and Sandberg, L. B., The synthesis of elastin, collagen, and glycosaminoglycans by high density primary cultures of neonatal rat aortic smooth muscle cells. An ultrastructural and biochemical study, *Eur. J. Cell Biol.,* 27, 34, 1982.
12. Hynes, R. O. and Yamada, K. M., Fibronectins: multifunctional modular glycoproteins, *J. Cell Biol.,* 95, 369, 1982.
13. Hynes, R. O., Fibronectin and its relation to cellular structure and behavior, in *Cell Biology of the Extracellular Matrix,* Hay, E. D., Ed., Plenum Press, New York, 1982, 295.
14. Mosher, D. F., Fibronectin, *Prog. Hemostasis Thromb.,* 5, 111, 1980.
15. Yamada, K. M., Cell surface interactions with extracellular materials, *Annu. Rev. Biochem.,* 52, 761, 1983.
16. Burke, J. M., Balian, G., Ross, R., and Bornstein, P., Synthesis of types I and III procollagen and collagen by monkey aortic smooth muscle cells *in vitro, Biochemistry,* 16, 3243, 1977.
17. Brennan, M. J., Millis, A. J. T., and Fritz, K. E., Fibronectin inhibits morphological changes in cultures of vascular smooth muscle cells, *J. Cell. Physiol.,* 112, 284, 1982.
18. Bornstein, P. and Ash, J. F., Cell surface-associated structural proteins in connective tissue cells, *Proc. Natl. Acad. Sci. U.S.A.,* 74, 2480, 1977.
19. Gold, L. I. and Pearlstein, E., Fibronectin-collagen binding and requirement during cellular adhesion, *Biochem. J.,* 186, 551, 1980.
20. Grotendorst, G. R., Seppa, H. E. J., Kleinman, H. K., and Martin, G. R., Attachment of smooth muscle cells to collagen and their migration towards platelet-derived growth factor, *Proc. Natl. Acad. Sci. U.S.A.,* 78, 3669, 1981.
21. Brennan, M. J., Millis, A. J. T., Mann, D., and Fritz, K. E., Structural alterations in fibronectin correlated with morphological changes in smooth muscle cells, *Dev. Biol.,* 97, 391, 1983.
22. Baenziger, N. L., Brodie, G. N., and Majerus, P. W., A thrombin-sensitive protein of human platelet membranes, *Proc. Natl. Acad. Sci. U.S.A.,* 68, 240, 1971.
23. Baenziger, N. L., Brodie, G. N., and Majerus, P. W., Isolation and properties of a thrombin-sensitive protein of human platelets, *J. Biol. Chem.,* 248, 2723, 1972.
24. McPherson, J., Sage, H., and Bornstein, P., Isolation and characterization of a glycoprotein secreted by aortic endothelial cells in culture: apparent identity with platelet thrombospondin, *J. Biol. Chem.,* 256, 11330, 1981.
25. Mosher, D. F., Doyle, M. F., and Jaffe, E. A., Synthesis and secretion of thrombospondin by cultured endothelial cells, *J. Cell Biol.,* 93, 343, 1982.
26. Raugi, G. J., Mumby, S. M., Abbott-Brown, D., and Bornstein, P., Thrombospondin: synthesis and secretion by cells in culture, *J. Cell Biol.,* 95, 351, 1982.

27. Jaffe, E. A., Ruggiero, J. T., Leung, L. L. K., Doyle, M. J., McKeown-Longo, P. J., and Mosher, D. F., Cultured human fibroblasts synthesize and secrete thrombospondin and incorporate it into extracellular matrix, *Proc. Natl. Acad. Sci. U.S.A.*, 80, 998, 1983.
28. Lawler, J. W., Slayter, H. S., and Coligan, J. E., Isolation and characterization of a high molecular weight glycoprotein from human blood platelets, *J. Biol. Chem.*, 253, 8609, 1978.
29. Margossian, S. S., Lawler, J. W., and Slayter, H. S., Physical characteristics of platelet thrombospondin, *J. Biol. Chem.*, 256, 7495, 1981.
30. Dixit, V. M., Grant, G. A., Santoro, S. A., and Frazier, W. A., Isolation and characterization of a heparin-binding domain from the amino terminus of platelet thrombospondin, *J. Biol. Chem.*, 259, 10100, 1984.
31. Raugi, G. J., Mumby, S. M., Ready, C. A., and Bornstein, P., Location and partial characterization of the heparin-binding fragment of platelet thrombospondin, *Thromb. Res.*, 36, 165, 1984.
32. Coligan, J. E. and Slayter, H. S., Structure of thrombospondin, *J. Biol. Chem.*, 259, 3944, 1984.
33. Mumby, S. M., Raugi, G. J., and Bornstein, P., Interactions of thrombospondin with extracellular matrix proteins: selective binding to type V collagen, *J. Cell Biol.*, 98, 646, 1984.
34. Lahav, J., Schwartz, M. A., and Hynes, R. O., Analysis of platelet adhesion with a radioactive chemical crosslinking reagent: interaction of thrombospondin with fibronectin and collagen, *Cell*, 31, 253, 1982.
35. Lahav, J., Lawler, J., and Gimbrone, M. A., Thrombospondin interactions with fibronectin and fibrinogen. Mutual inhibition in binding, *Eur. J. Biochem.*, 145, 151, 1984.
36. Jaffe, E. A., Leung, L. L. K., Nachman, R. L., Levin, R. I., and Mosher, D. F., Thrombospondin is the endogenous lectin of human platelets, *Nature (London)*, 295, 246, 1982.
37. Leung, L. L. K., Role of thrombospondin in platelet aggregation, *J. Clin. Invest.*, 74, 1764, 1984.
38. Mumby, S. M., Abbott-Brown, D., Raugi, G. J., and Bornstein, P., Regulation of thrombospondin secretion by cells in culture, *J. Cell. Physiol.*, 120, 280, 1984.
39. Majack, R. A., Cook, S. C., and Bornstein, P., Platelet-derived growth factor and heparin-like glycosaminoglycans regulate thrombospondin synthesis and deposition in the matrix by smooth muscle cells, *J. Cell Biol.*, 101, 1059, 1985.
40. Burke, J. M. and Ross, R., Synthesis of connective tissue molecules by smooth muscle, *Int. Rev. Connect. Tissue Res.*, 8, 119, 1979.
41. Foster, J. A., Elastin structure and biosynthesis: an overview, *Methods Enzymol.*, 82, 559, 1982.
42. Franzblau, C. and Faris, B., Biosynthesis of insoluble elastin in cell and organ cultures, *Methods Enzymol.*, 82, 615, 1982.
43. Eyre, D. W., Paz, M. A., and Gallop, P. M., Crosslinking in collagen and elastin, *Annu. Rev. Biochem.*, 53, 717, 1984.
44. Dunn, D. M. and Franzblau, C., Effects of ascorbate on insoluble elastin accumulation and cross-link formation in rabbit pulmonary artery smooth muscle cell cultures, *Biochemistry*, 21, 4195, 1982.
45. Faris, B., Salcedo, L. L., Cook, V., Johnson, L., Foster, J. A., and Franzblau, C., The synthesis of connective tissue protein in smooth muscle cells, *Biochim. Biophys. Acta*, 418, 93, 1976.
46. Snider, R., Faris, B., Verbitzski, V., Moscaritolo, R., Salcedo, L., and Franzblau, C., Elastin biosynthesis and cross-link formation in rabbit aortic smooth muscle cell cultures, *Biochemistry*, 20, 2614, 1981.
47. Foster, J. A., Mecham, R. P., Rich, C. B., Cronin, M. F., Levine, A., Imberman, M., and Salcedo, L. L., Proelastin. Synthesis in cultured smooth muscle cells, *J. Biol. Chem.*, 253, 2797, 1978.
48. Abraham, P. A., Hart, M. L., Winge, A. R., and Carnes, W. H., The biosynthesis of elastin by an aortic medial cell culture, *Adv. Exp. Med. Biol.*, 79, 397, 1977.
49. Narayanan, A. S., Sandberg, L. B., Ross, R., and Layman, D. L., The smooth muscle cell. VI. Elastin synthesis in arterial smooth muscle cell culture, *J. Cell Biol.*, 68, 411, 1976.
50. Ross, R., The smooth muscle cell. II. Growth of smooth muscle in culture and formation of elastic fibers, *J. Cell Biol.*, 50, 172, 1971.
51. Hinek, A. and Thyberg, J., Electron microscopic observations on the formation of elastic fibers in primary cultures of aortic smooth muscle cells, *J. Ultrastruct. Res.*, 60, 12, 1977.
52. Muir, L. W., Bornstein, P., and Ross, R., A presumptive subunit of elastic fiber microfibrils secreted by arterial smooth muscle cells in culture, *Eur. J. Biochem.*, 64, 105, 1976.
53. Rucker, R. B. and Tinker, D., Structure and metabolism of arterial elastin, *Int. Rev. Exp. Pathol.*, 17, 1, 1977.
54. Cleary, E. G. and Gibson, M. A., Elastin-associated microfibrils and microfibrillar proteins, *Int. Rev. Connect. Tissue Res.*, 10, 97, 1983.
55. Mayne, R., Vascular connective tissue: normal biology and derangement in human diseases, in *Diseases of Connective Tissue: The Molecular Pathology of the Extracellular Matrix*, Uitto, J. and Perejda, Z., Eds., Marcel Dekker, New York, 1984.

56. Leung, D. Y. M., Glagov, S., and Mathews, M. B., A new *in vitro* system for studying cell response to mechanical stimulation, *Exp. Cell Res.*, 109, 285, 1977.
57. Majack, R. A. and Bornstein, P., Heparin regulates the collagen phenotype of vascular smooth muscle cells: induced synthesis of an M_r 60,000 collagen, *J. Cell Biol.*, 100, 613, 1985.
58. Majack, R. A. and Bornstein, P., Regulation of collagen biosynthesis: heparin alters the biosynthetic phenotype of vascular smooth muscle cells in culture, *Ann. N.Y. Acad. Sci.*, 460, 172, 1985.
59. Beldekas, J. C., Smith, B., Gerstenfeld, L. C., Sonenshein, G. E., and Franzblau, C., Effects of 17β-estradiol on the biosynthesis of collagen in cultured bovine aortic smooth muscle cells, *Biochemistry*, 20, 2162, 1981.
60. Pietila, K. and Nikkari, T., Enhanced synthesis of collagen and total protein by smooth muscle cells from atherosclerotic rabbit aortas in culture, *Atherosclerosis*, 37, 11, 1980.
61. Stavenow, L., Kjellstrom, T., and Malmquist, J., Stimulation of collagen production in growth-arrested myocytes and fibroblasts in culture by growth factors from platelets, *Exp. Cell Res.*, 136, 321, 1981.
62. Bornstein, P. and Sage, H., Structurally distinct collagen types, *Annu. Rev. Biochem.*, 49, 957, 1980.
63. Holderbaum, D. and Ehrhart, L. A., Modulation of types I and III procollagen synthesis at various stages of arterial smooth muscle cell growth *in vitro*, *Exp. Cell Res.*, 153, 16, 1984.
64. Scott, D. M., Harwood, R., Grant, M. E., and Jackson, D. S., Characterization of the major collagen species present in porcine aortae and the synthesis of their precursors by smooth muscle cells in culture, *Connect. Tissue Res.*, 5, 7, 1977.
65. Rauterberg, J., Allam, S., Brehmer, U., Wirth, W., and Hauss, W. H., Characterization of the collagen synthesized by cultured human smooth muscle cells from fetal and adult aorta, *Hoppe Seyler's Z. Physiol. Chem.*, 358, 401, 1977.
66. Layman, D. L., Epstein, E. H., Dodson, R. F., and Titus, J. L., Biosynthesis of type I and III collagens by cultured smooth muscle cells from human aorta, *Proc. Natl. Acad. Sci. U.S.A.*, 74, 671, 1977.
67. Lohler, J., Timpl, R., and Jaenisch, R., Embryonic lethal mutation in mouse collagen I gene causes rupture of blood vessels and is associated with erythropoietic and mesenchymal cell death, *Cell*, 38, 597, 1984.
68. Beldekas, J. C., Gerstenfeld, L., Sonenshein, G. E., and Franzblau, C., Cell denstiy and estradiol modulation of procollagen type III in cultured calf smooth muscle cells, *J. Biol. Chem.*, 251, 12252, 1982.
69. Gerstenfeld, L., Beldekas, J. C., Sonenshein, G. E., and Franzblau, C., Processing of procollagen types III and I in cultured bovine smooth muscle cells, *J. Biol. Chem.*, 259, 9158, 1984.
70. Byers, P. H., Holbrook, K. A., McGillivray, B., MacLeod, P. M., and Lowry, R. B., Clinical and structural heterogeneity of type IV Ehlers-Danlos syndrome, *Hum. Genet.*, 47, 141, 1979.
71. Majack, R. A. and Bornstein, P., Heparin and related glycosaminoglycans regulate the secretory phenotype of vascular smooth muscle cells, *J. Cell Biol.*, 99, 1688, 1984.
72. Mayne, R., Vail, M. S., Miller, E. J., Blose, S. H., and Chacko, S., Collagen polymorphism in cell cultures derived from guinea pig aortic smooth muscle: comparison with three populations of fibroblasts, *Arch. Biochem. Biophys.*, 181, 462, 1977.
73. Aumailley, M. and Bricaud, H., Collagen synthesis in organ culture of normal and atherosclerotic aortas, *Atherosclerosis*, 39, 1, 1981.
74. MacCallaugh, K. A. and Balian, G., Collagen characterization and cell transformation in human atherosclerosis, *Nature (London)*, 258, 73, 1975.
75. Morton, L. F. and Barnes, M. J., Collagen polymorphism in the normal and diseased blood vessel wall, *Atherosclerosis*, 42, 41, 1982.
76. Odermatt, B. F., Lang, A. B., Ruttner, J. R., Winterhalter, K. H., and Trueb, B., Monoclonal antibodies to human type IV collagen: useful reagents to demonstrate the heterotrimeric nature of the molecule, *Proc. Natl. Acad. Sci. U.S.A.*, 81, 7343, 1984.
77. Timpl, R. and Martin, G. R., Components of basement membranes, in *Immunochemistry of the Extracellular Matrix*, Vol. 2, Furthmayr, H., Ed., CRC Press, Boca Raton, Fla., 1982, 119.
78. Sankey, E. A. and Barnes, M. J., Comparison of the collagenous products synthesized in culture by pig aortic endothelial and smooth muscle cells, *Biochem. J.*, 218, 11, 1984.
79. Fessler, J. H. Shigaki, N., and Fessler, L. I., Biosynthesis and properties of type V procollagens, *Ann. N. Y. Acad. Sci.*, 460, 181, 1985.
80. Niyibizi, C., Fietzek, P. P., and van der Rest, M., Human placenta type V collagens. Evidence for the existence of an $\alpha 1(V)\alpha 2(V)\alpha 3(V)$ collagen molecule, *J. Biol. Chem.*, 259, 14170, 1984.
81. Mayne, R., Vail, M. S., and Miller, E. J., Characterization of the collagen chains synthesized by cultured smooth muscle cells derived from rhesus monkey thoracic aorta, *Biochemistry*, 17, 446, 1978.

82. Gay, S., Martinez-Hernandez, A., Rhodes, R. K., and Miller, E. J., The collagenous exocytoskeleton of smooth muscle cells, *Collagen Rel. Res.*, 1, 377, 1981.
83. Chung, E., Rhodes, R. K., and Miller, E. J., Isolation of three collagenous components of probable basement membrane origin from several tissues, *Biochem. Biophys. Res. Commun.*, 71, 1167, 1976.
84. Odermatt, E., Risteli, J., Van Delden, V., and Timpl, R., Structural diversity and domain composition of a unique collagenous fragment (intima collagen) obtained from human placenta, *Biochem. J.*, 211, 295, 1983.
85. Jander, R., Rauterberg, J., and Glanville, R. W., Further characterization of the three polypeptide chains of bovine and human short-chain collagen (intima collagen), *Eur. J. Biochem.*, 133, 39, 1983.
86. Jander, R., Troyer, D., and Rauterberg, J., A collagen like glycoprotein of the extracellular matrix is the undegraded form of type VI collagen, *Biochemistry*, 23, 3675, 1984.
87. Trueb, B. and Bornstein, P., Characterization of the precursor form of type VI collagen, *J. Biol. Chem.*, 259, 8597, 1984.
88. Hessle, H. and Engvall, E., Type VI collagen. Studies of its localization, structure and biosynthetic form with monoclonal antibodies, *J. Biol. Chem.*, 259, 3955, 1984.
89. von der Mark, H., Aumailley, M., Wick, G., Fleischmajer, R., and Timpl, R., Immunochemistry, genuine size and tissue localization of collagen VI, *Eur. J. Biochem.*, 142, 493, 1984.
90. Schmid, T. M. and Conrad, H. E., A unique low molecular weight collagen secreted by cultured chick embryo chondrocytes, *J. Biol. Chem.*, 257, 12444, 1982.
91. Gibson, G. J., Schor, S. L., and Grant, M. E., Effects of matrix molecules on chondrocyte gene expression: synthesis of a low molecular weight collagen species by cells cultured within collagen gels, *J. Cell Biol.*, 93, 767, 1982.
92. Schmid, T. M. and Linsenmayer, T. F., Localization of short chain skeletal collagen, *Fed. Proc.*, 43, 1588, 1984.
93. Ross, R. and Vogel, A., The platelet-derived growth factor, *Cell*, 14, 203, 1978.
94. Bernstein, L. R., Antoniades, H., and Zetter, B. R., Migration of cultured vascular cells in response to plasma and platelet-derived factors, *J. Cell Sci.*, 56, 1, 1982.
95. Nakao, J., Ooyama, T., Chang, W. C., Murota, S., and Orimo, H., Platelets stimulate aortic smooth muscle cell migration, *in vitro, Atherosclerosis*, 43, 143, 1982.
96. Grotendorst, G. R., Chang, T., Seppa, H. E. J., Kleinman, H. K., and Martin, G. R., Platelet-derived growth factor is a chemoattractant for vascular smooth muscle cells, *J. Cell. Physiol.*, 113, 261, 1983.
97. Clowes, A. W., Clowes, M. M., Gown, A. M., and Wight, T. N., Localization of proteoheparan sulfate in rat aorta, *Histochemistry*, 80, 379, 1984.
98. Marcum, J. A. and Rosenberg, R. D., Anticoagulantly active heparin-like molecules from vascular tissue, *Biochemistry*, 23, 1730, 1984.
99. Ross, R., Vogel, A., Davies, P., Raines, E., Kariya, B., Rivest, M. J., Gustafson, C., and Glomset, J., The platelet derived growth factor and plasma control cell proliferation, in *Hormones and Cell Culture*, Book A, Sato, G. and Ross, R., Eds., Cold Spring Harbor Press, Cold Spring Harbor, N.Y., 1979, 3.
100. Stiles, C. D., Pledger, W. J., Tucker, R. W., Martin, R. G., and Scher, C. D., Regulation of the Balb/c-3T3 cell cycle — effects of growth factors, *J. Supramol. Struct.*, 13, 489, 1980.
101. Scher, C. D., Shepard, R. C., Antoniades, H. N., and Stiles, C. D., Platelet-derived growth factor and the regulation of the mammalian fibroblast cell cycle, *Biochim. Biophys. Acta*, 560, 217, 1979.
102. Cochran, B. H., Lillquist, J. S., and Stiles, C. D., Post-transcriptional control of protein synthesis in Balb/c-3T3 cells by platelet-derived growth factor and platelet-poor plasma, *J. Cell. Physiol.*, 109, 429, 1981.
103. Burke, J. M. and Ross, R., Collagen synthesis by monkey arterial smooth muscle cells during proliferation and quiescence in culture, *Exp. Cell Res.*, 107, 387, 1977.
104. Cochran, B. H., Reffel, A. C., and Stiles, C. D., Molecular cloning of gene sequences regulated by platelet-derived growth factor, *Cell*, 33, 939, 1983.
105. Hendrickson, S. L. and Scher, C. D., Platelet-derived growth factor modulated translatable mRNAs, *Mol. Cell. Biol.*, 3, 1478, 1983.
106. Pledger, W. J., Hart, C. A., Locatell, K. L., and Scher, C. D., Platelet-derived growth factor-modulated proteins: constitutive synthesis by a transformed cell line, *Proc. Natl. Acad. Sci. U.S.A.*, 79, 4358, 1981.
107. Greenburg, M. E. and Ziff, E. B., Stimulation of 3T3 cells induces transcription of the c-fos proto-oncogene, *Nature (London)*, 311, 433, 1983.
108. Kelly, K., Cochran, B. H., Stiles, C. D., and Leder, P., Cell-specific regulation of the c-myc gene by lymphocyte mitogens and platelet-derived growth factor, *Cell*, 35, 603, 1983.

109. Nilsen-Hamilton, M., Shapiro, J. M., Massoglia, S. L., and Hamilton, R. T., Selective stimulation by mitogens of incorporation of [^{35}S]-methionine into a family of proteins released into the medium by 3T3 cells, *Cell*, 20, 19, 1980.
110. Nilsen-Hamilton, M., Hamilton, R. T., and Adams, G. A., Rapid selective stimulation by growth factors of the incorporation by Balb/c-3T3 cells of (^{35}S)-methionine into a glycoprotein and 5 super-inducible proteins, *Biochem. Biophys. Res. Commun.*, 108, 158, 1982.
111. Gospodarowicz, D. and Ill, C. R., Do plasma and serum have different abilities to promote cell growth? *Proc. Natl. Acad, Sci. U.S.A.*, 77, 2726, 1980.
112. Gospodarowicz, D., Vlodavsky, I., and Savion, N., The extracellular matrix and the control of proliferation of vascular endothelial and vascular smooth muscle cells, *J. Supramol. Struct.*, 13, 339, 1980.
113. Libby, P. and O'Brien, K. V., Culture of quiescent arterial smooth muscle cells in a defined serum-free medium, *J. Cell. Physiol.*, 115, 217, 1983.
114. Salomon, D. S., Liotta, L. A., and Kidwell, W. R., Differential response to growth factor by rat mammary epithelium plated on different collagen substrata in serum-free medium, *Proc. Natl. Acad. Sci. U.S.A.*, 78, 382, 1981.
115. Bitterman, P. B., Rennard, S. I., Adelberg, S., and Crystal, R. G., Role of fibronectin as a growth factor for fibroblasts, *J. Cell Biol.*, 97, 1925, 1983.
116. Wicha, M. S., Liotta, L. A., Garbisa, S., and Kidwell, W. R., Basement membrane collagen requirements for attachment and growth of mammary epithelium, *Exp. Cell Res.*, 124, 181, 1979.
117. Tseng, S. C. G., Savion, N., Stern, R., and Gospodarowicz, D., Fibroblast growth factor modulates synthesis of collagen in cultured vascular endothelial cells, *Eur. J. Biochem.*, 122, 355, 1982.
118. Hata, R., Hori, H., Nagai, Y., Tanaka, S., Kondo, M., Hiramatsu, M., Utsum, N., and Kumegawa, M., Selective inhibition of type I collagen synthesis in osteoblastic cells by epidermal growth factor, *Endocrinology*, 115, 867, 1984.
119. Steinmann, B. U., Abe, S., and Martin, G. R., Modulation of type I and III collagen production in normal and mutant human skin fibroblasts by cell density, prostaglandin E_2 and epidermal growth factor, *Collagen Rel. Res.*, 2, 185, 1982.
120. Clowes, A. W. and Karnovsky, M. J., Suppression by heparin of smooth muscle cell proliferation in injured arteries, *Nature (London)*, 265, 625, 1977.
121. Guyton, J. R., Rosenberg, R. D., Clowes, A. W., and Karnovsky, M. J., Inhibition of rat arterial smooth muscle cell proliferation by heparin. I. *In vivo* studies with anticoagulant and non-anticoagulant heparin, *Circ. Res.*, 46, 625, 1980.
122. Hatcher, V. B., Tsien, G., Oberman, M. S., and Burk, P. G., Inhibition of cell proliferation and protease activity by cartilage factors and heparin, *J. Supramol. Struct.*, 14, 33, 1980.
123. Nilsson, J., Ksiazek, T., Thyberg, J., and Wasteson, A., Cell surface components and growth regulation in cultivated arterial smooth muscle cells, *J. Cell Sci.*, 64, 107, 1983.
124. Hoover, R. L., Rosenberg, R. D., Haering, W., and Karnovsky, M. J., Inhibition of rat arterial smooth muscle cell proliferation by heparin. II. *In vitro* studies, *Circ. Res.*, 47, 578, 1980.
125. Gamse, G., Fromme, H. G., and Kresse, H., Metabolism of sulfated glycosaminoglycans in cultured endothelial cells and smooth muscle cells from bovine aorta, *Biochim. Biophys. Acta*, 544, 514, 1978.
126. Castellot, J. J., Addonizio, M. L., Rosenberg, R. D., and Karnovsky, M. J., Cultured endothelial cells produce a heparinlike inhibitor of smooth muscle cell growth, *J. Cell Biol.*, 90, 372, 1981.
127. Castellot, J. J., Favreau, L. V., Karnovsky, M. J., and Rosenberg, R. D., Inhibition of vascular smooth muscle cell growth by endothelial cell-derived heparin, *J. Biol. Chem.*, 257, 11256, 1982.
128. Willems, C., Astaldi, G. C. B., de Groot, P. G., Janssen, M. C., Gonsalvez, M. D., Zeijlemaker, W. P., van Mourik, J. A., and van Aken, W. G., Media conditioned by cultured human vascular endothelial cells inhibit the growth of vascular smooth muscle cells, *Exp. Cell Res.*, 139, 191, 1981.
129. Oosta, G. M., Favreau, L. V., Beeler, D. L., and Rosenberg, R. D., Purification and properties of human platelet heparitinase, *J. Biol. Chem.*, 257, 11249, 1982.
130. Fritze, L. M. S. and Rosenberg, R. D., Bovine aortic smooth muscle cells produce an inhibitor of smooth muscle cell growth *in vitro*, *J. Cell Biol.*, 97, 90a, 1983.
131. Wight, T. N. and Hascall, V. C., Proteoglycans in primate arteries. III. Characterization of the proteoglycans synthesized by arterial smooth muscle cells in culture, *J. Cell Biol.*, 96, 167, 1983.
132. Majack, R. A. and Clowes, A. W., Inhibition of vascular smooth muscle cell migration by heparin-like glycosaminoglycans, *J. Cell. Physiol.*, 118, 253, 1984.
133. Millis, A. J. T., Hoyle, M., and Reich, E., Nodular smooth muscle cells secrete a unique heparin-binding protein, *J. Cell Biol.*, 99, 29a, 1984.
134. Majack, R. A., unpublished observations.

Chapter 9

SYNTHESIS OF GLYCOSAMINOGLYCANS AND PROTEOGLYCANS

M. J. Merrilees

TABLE OF CONTENTS

I. INTRODUCTION

Vascular smooth muscle cells (SMC) synthesize and secrete proteoglycans (PG) and constituent glycosaminoglycans (GAG), matrix components which together with collagen and elastin play a vital role in maintaining the structural integrity of vessel walls. Interest in factors and mechanisms controlling the production of PG and GAG by SMC stems in large part from observations that the amounts and proportions of the various types of GAG are changed at sites of atherosclerotic lesion formation[1,2] and from findings that sulphated GAG (SGAG) have the ability to form complexes with low- and very low-density lipoproteins.[3-7] These two features have provided the impetus in recent years for a large number of investigations on the effects of atherosclerotic risk factors on GAG synthesis, and many of these studies have been conducted in vitro, using cell and organ culture techniques.

Like other connective tissue cells in culture SMC synthesize a spectrum of GAG including hyaluronic acid (HA), heparan sulfate (HS), dermatan sulfate (DS), and chondroitin sulfates A (CSA or CS_4) and C (CSC or CS_6). These GAG, except for HA, do not exist as free chains but are covalently bound to protein to form PG macromolecules, which may further link with HA to form very large PG aggregates.[8] These PG and PG aggregates are the functional components of the matrix, but, interestingly, there have been comparatively few studies on PG synthesis by SMC in culture. Most investigators have preferred to analyze the constituent GAG. While PG production is currently receiving increased attention,[9,10,98] analyses of GAG are nevertheless of considerable value since the amounts and proportions of GAG are very good indicators to the mechanical properties of tissues.[11] The amounts and proportions of GAG produced by SMC under conditions of cell and organ culture, however, are highly variable, and the interpretation of results from in vitro studies and their extrapolation to in vivo requires considerable caution. This chapter will review the effects of those factors which contribute to this variability in GAG synthetic patterns in culture and, further, will review the effects on GAG production of a number of factors considered to play a role in atherogenesis.

II. FACTORS AFFECTING GLYCOSAMINOGLYCAN LEVELS IN CULTURE

A. Culture Conditions

1. Cell Density and Phase of Growth

Changes in the amounts and proportions of GAG with changes in cell density probably best illustrate the importance of carefully monitoring culture conditions during experimental work. The relationship is complex and, surprisingly, only a few papers have reported in detail on the effects of cell density and phase of growth on SMC GAG synthesis.

Immediately after seeding at a low density, SMC are particularly active in the synthesis of GAG,[12] and HA is usually the predominate component.[12,13] This high level of activity is not solely due to increased synthesis associated with initial attachment and cell migration. Wight and Ross[14] found that cultures of primate SMC, maintained at a low density for an extended period by culturing initially in 1% serum for 7 days followed by serum-free medium for 12 days, also displayed a high level of GAG synthesis compared with more dense cultures.

The exponential or proliferative phase of growth, whether occurring in the first few days after seeding[12] or initiated at a later stage by increasing the serum level,[14] is associated with a marked reduction in GAG synthesis compared to low-density cultures

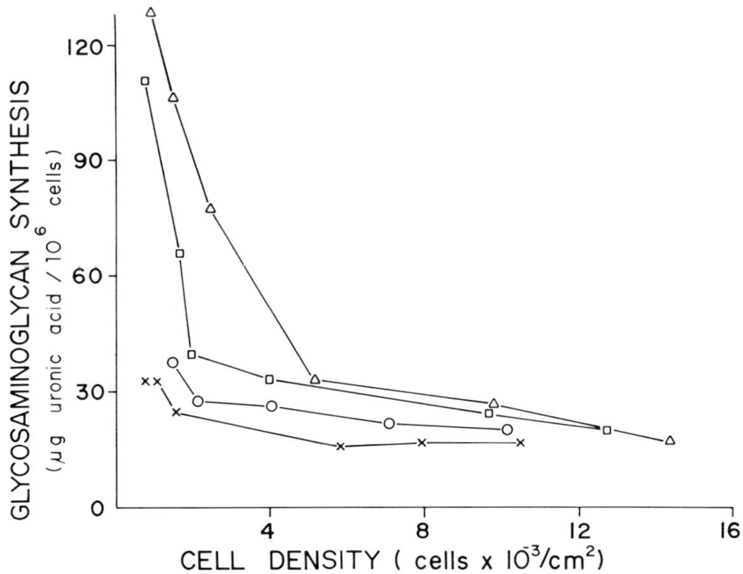

FIGURE 1. Effect of serum concentration and cell density on GAG production by SMC in culture. Cells were plated out at six different densities from 9 to 100×10^2 cells per cm^2 and when established given a serum-free medium for 24 hr. Subgroups were then given medium containing 0% (×), 1% (O), 5% (□), or 10% (△) serum. After a further 24 hr, cell counts were made and medium assayed for total GAG. (From Wosu, L., Parisella, R., and Kalant, N., *Atherosclerosis*, 48, 205, 1983. With permission.)

(Figure 1). Furthermore, during the proliferation phase and as cultures approach the stationary phase, the proportion of HA decreases while that of the SGAG increases.[12,13,15,16]

Few studies, however, have distinguished between the effect of cell density per se and the effect of proliferation rate. Wosu et al.[17] recently examined this feature and concluded that the fall in the rate of GAG synthesis which occurs with increasing cell density was a direct effect of density and not related to proliferation rate which was biphasic.

The reduced rate of synthesis in high-density stable cultures is not necessarily a constant feature. Pietilä[12] measured GAG synthesis in cultures of rabbit SMC maintained from 5 to 16 days in a stationary phase at high density and found an increase in GAG synthesis, including an increase in HA, compared to the proliferative and early stationary phases. The rates of synthesis, however, were not as high as that found for the low-density cultures immediately after seeding. Others have found that while SMC maintained at high densities for prolonged periods are still active in GAG synthesis, only very small amounts of HA are detectable,[14,18] and the proportions of GAG in these cultures more closely resemble the proportions in vivo.

2. Serum

Serum concentration is positively correlated with cell density, and the effects of serum on GAG synthesis can in part be explained on the basis of changes in cell densities. Serum does, however, affect GAG synthesis independently of cell density. Wosu et al.[17] showed that total GAG synthesis by human SMC increased with increasing serum concentration (0 to 10%), although at higher densities (where GAG synthesis was reduced) the effect was much less marked (Figure 1). Both SGAG and HA are stimulated by serum. The response of SGAG to changed serum levels, however, is more complicated than that for HA. Pietila[12] found that reduction in the serum level of cultures maintained in 10% serum resulted initially (24 to 48 hr) in an increase in SGAG synthesis, but after 7 days in the appropriate serum concentration, SGAG synthesis was positively correlated with serum levels. In contrast, HA was immediately responsive and positively correlated with serum concentration irrespective of the length of time of preincubation with serum. HA also shows a greater response to serum than SGAG,[16] which may relate to the finding that serum contains a factor capable of stimulating hyaluronic acid synthetase activity in fibroblasts.[19]

3. pH

Pietilä[12] examined the effect of pH, ranging from 6.8 to 7.9, on rabbit SMC cultured both in the presence of 10% serum and in serum-free medium. Synthesis of both SGAG and HA was shown to be positively correlated with pH and over the range tested there was a seven- to tenfold difference in incorporation irrespective of serum levels.

Other factors such as hypoxia and additives such as ascorbic acid also affect GAG synthesis, but these will be considered under other headings later. Cell density, serum, and pH are variables encountered under standard culture conditions, and it is evident that alterations in any one of these factors results in marked changes to both the amounts and proportions of GAG.

B. Source of SMC

1. Vessels

Almost all SMC cultures for the study of GAG metabolism have been derived from thoracic aortas, and most cultures have been established from medial or intimal-medial explants. Where SMC derived from intima have been compared to SMC from the media of the same animal, the patterns of GAG synthesis have been found to be very similar, apart from some minor differences in the pericellular pools.[20] Interestingly, muscular arteries, such as the coronary arteries where atherosclerotic plaques of primary significance occur, have rarely been used. Wosu et al.[17] used SMC from adult human tibial arteries (muscular) and from fetal human aortas in their study on the effect of lipoprotein on GAG secretion, but a direct comparison between the two types of SMC was not made. Thus it remains to be determined whether or not SMC cultured from different types of vessels have similar or different GAG synthetic patterns.

SMC from the same vessel of the same species, however, can display quite marked differences in the amounts and proportions of GAG, and these differences relate to the state of the vessel wall from which the SMC cultures were derived. Pietila et al.[21] examined GAG synthesis by rabbit SMC, established from intimal-medial explants, from normal vessels, and from vessels made atherosclerotic by 6 months of cholesterol feeding. They found that SMC derived from the thickened atherosclerotic vessels had a significantly higher rate of GAG synthesis than normal cells and that most of the increase was due to an increase in SGAG, predominately DS. This finding, however, was a feature only of proliferative lesions. In more advanced stable fibrous lesions, formed by 6 months of cholesterol feeding followed by 6 months on a normal diet, the synthesis of sulfated GAG was slightly but significantly reduced compared to control SMC, while the synthesis of HA was increased.[22]

The synthesis of SGAG by SMC derived from atherosclerotic-susceptible White Carneau (WC) pigeons is also greater than for SMC from the atherosclerotic-resistant Show Racer (SR) breed. SMC from WC pigeons synthesized three to four times as much SGAG as SR pigeons, although the relative proportions of the individual SGAG remained the same.[23] Another example of intraspecific variation in GAG synthesis is from SMC cultured from hypertensive rats. Their SMC synthesize more SGAG than SMC from normotensive rats, although this difference can be related directly to the higher average protein content of SMC from hypertensive rats.[24]

2. Species

The percentage distribution of the individual GAG varies considerably between species (Table 1). Not all of this variation, however, is attributable to true differences between species. Some of the variation is clearly due to culture conditions while differences in methods of GAG analysis may also be a factor.

Differences in methodology probably contribute only a small amount to the variation. In the majority of studies the separation of individual GAG has been by similar methods: protease digestion, ethanol or quarternary ammonium salt precipitation, followed by electrophoresis. Methods for the isolation and analysis of GAG were recently reviewed by Taniguchi,[25] and electrophoretic techniques were considered to give good separation of most GAG, recognizing that GAG display appreciable heterogeneity including differences in degree of sulfation, polydispersity of molecular weight, and variable amounts of attached residual peptides. Problems with methodology are partly resolved where comparative studies have been undertaken in the same laboratory using a standardized method. Under these circumstances, it would appear that at least some of the apparent species variability can be explained by variations in culture conditions, particularly cell density.

The major apparent variation between species relates to the amount of HA. Confluent cultures of nonhuman primates are characterized generally by a low percentage (0 to 10%) of HA, while confluent SMC from adult and fetal humans, pigs, and rabbits have, in general, a high proportion of HA, frequently more than 50% of the total GAG. Unfortunately, in most studies cell densities are not specified but only reported as being confluent, subconfluent, or in the growth phase. Where cell densities have been specified, or where stationary and growth phase cultures have been compared directly, it is apparent that low-density cultures are usually characterized by a high proportion of HA. Thus rat SMC, for example, produce mostly HA (70%) at low densities (0.6×10^6 cells per 25 cm^2) but only a small proportion (10%) at a high densities (3×10^6 cells per 25 cm^2).[16] If cultured for an extended period of up to 4 weeks, during which time cells are allowed to accumulate and form multilayers, only a trace of HA is detectable.[18] A similar reduction in the proportion of HA with increasing density has been found for porcine SMC,[13] but it would appear that human SMC, even when confluent, still produce mostly HA (see Table 1).

The proportions of the SGAG are also variable but unlike HA not so clearly correlated with changes in cell density. Dense cultures generally show an increased proportion of DS compared to low-density cultures, but there is no consistent pattern of dominance by either DS or CS with respect to density. Rather there seems to be a species-specific pattern for the SGAG, and this pattern correlates in most instances with the pattern occurring in vivo. For example, in some nonhuman primates and in rats, DS is dominant both in vivo and in vitro while in pigs, bovines, pigeons, and humans, CS is dominant in both situations. There are, however, a number of inconsistencies with this correlation. Very dense long-term cultures of rat SMC[18] have a predominance of CS (43%), although DS still constitutes a significant proportion (32%). A similar distribution has been found in some low-density cultures of rat SMC.[26] Chondroitin sulfate is dominant in rabbit aorta,[27,28] but in culture, rabbit SMC usually synthesize as much or more DS.[29,30]

Heparan sulfate constitutes on average about 20 to 25% of the total GAG in the intima and media, and, as for the other SGAG, there is species variability. Rat aorta for example is especially rich in HS (40 to 50%),[27,28,31] while the proportion in pigeon

Table 1
PERCENTAGE DISTRIBUTION OF GLYCOSAMINOGLYCANS PRODUCED BY SMC

Species[a]	Cell density	Serum level (%)	Distribution of GAG HA	HS	DS	CS A/C	Compartment	Ref.
Human (fetal)	Confluent	15	71	7	3	19	Medium	60
Human (fetal)	Confluent	15	50	(	50	)	Total GAG	48
Human (fetal)	Confluent	10	60	(	40	)	Total GAG	61
Human (fetal)	Confluent	15	71	(	29	)	Medium	50
			45	(	55	)	Cell layer[b]	
			67	(	33	)	Total GAG	
Human (fetal)	Confluent	LPF[c]						
		5	65	(	35	)	Total GAG	17
		2	70	(	30	)	Total GAG	
		LDL[d]						
		5	56	(	44	)	Total GAG	
		2	61	(	39	)	Total GAG	
Pigtail monkey	Confluent	5	0—5	ND[e]	80	15	Medium	14
Cynomolgus monkey	Organ culture	0	9	12	27	53	Whole vessel wall	42
Bovine	Subconfluent	0	5	2	2	91	Medium	98
			14	10	10	66	Cell Layer	
Bovine (fetal)		10	0	60	8	32	Pericellular coat	95
Bovine	Confluent	10	ND[e]	22	20	58	Medium	20
Bovine	Organ culture		30	11	10	49	Whole vessel wall	100
Porcine	Subconfluent	2	49	13	18	20	Medium	13
	Confluent for 10 days	2	18	17	31	34	Medium	
Porcine	Confluent	10	20	10	15	55	Medium	43
			17	40	(43)		Pericellular coat	
Porcine	Confluent	2	24	9	20	47	Total GAG	26
Porcine	Confluent	10	23	10	(67)		Medium	9
Porcine	Confluent	1	25	10	30	35	Total GAG	78
Rabbit (weanlings)	Confluent	10	64	2	23	11	Medium	30
	$1—1.5 \times 10^6$[f]	10	13	31	40	16	Cell Layer	
Rabbit	Confluent	10	62	(	38	)	Total GAG	21
Rabbit	Subconfluent	0	57	(	43	)	Total GAG	12
	Confluent	10	89	(	11	)	Total GAG	
Rabbit	Confluent	10	63	(	37	)	Total GAG	22
Rat (Neonatal)	Confluent multilayers	20	Trace	30	30	40	Medium	18
Rat	Subconfluent	2	24	8	21	47	Total GAG	26
	Confluent	2	9	11	50	30	Total GAG	
Rat	Subconfluent (lag phase)	10	70	13	12	5	Intracellular 1° cultures	15
	Confluent (stationary phase)	10	9	34	55	2	Intracellular 1° cultures	
	Confluent	10	10	59	24	7	Pericellular coat 1° culture	
	Confluent	10	3	7	88	2	Intracellular S10	
	Confluent	10	19	37	42	2	Pericellular coat S10	
Rat	3.0×10^6	1	10	10	43	37	Medium	16
	2.0×10^6	1	18	9	61	12	Medium	
	1.5×10^6	1	28	5	41	26	Medium	
	0.6×10^6	1	72	5	13	10	Medium	
	2.0×10^6	1	56	23	15	6	Cell layer	
	1.5×10^6	1	45	26	11	18	Cell layer	

Table 1 (continued)
PERCENTAGE DISTRIBUTION OF GLYCOSAMINOGLYCANS PRODUCED BY SMC

Species[a]	Cell density	Serum level (%)	Distribution of GAG HA	HS	DS	CS A/C	Compartment	Ref.
Pigeon								
White Carneau	Confluent	15	ND	5	35	60	Medium	23
Show Racer	Confluent	15	ND	5	35	60	Medium	

[a] All SMC for all species are from aortas.
[b] Cells plus pericellular coat.
[c] Lipoprotein-free serum.
[d] Low-density lipoprotein-containing serum.
[e] Not determined.
[f] Cell density per 25 cm² flask.

aorta may be as low as 11%.[32] Electron microscopical[33] and immunohistochemical[34] analyses suggest that HS is located primarily in the basement membrane-like material which surrounds individual SMC in the arterial walls (see Chapters 4 and 7). This basement membrane-like material has been isolated from rabbit SMC in culture and shown to contain predominantly HS (55%) and chondroitin sulfate with a small amount of HA and a trace of DS.[35] HS and CS also predominate in the pericellular coat immediately adjacent to the plasma membrane of cultured SMC.[36] In other culture studies, where the cell layer and growth medium have been analyzed separately, the proportion of HS is markedly increased in the pericellular pool (Table 1), a feature characteristic of other cell types in culture.[37-39]

C. SMC Phenotype

Variability in the amounts and proportions of GAG synthesized by SMC in culture may also be due partly to variability in SMC phenotype. The vast majority of studies, however, have not taken into account the degree of modulation of SMC which may occur in culture (see Chapter 4). Recently, Yamamoto et al.[15] did address this point directly by measuring GAG synthesis by rat SMC as the cells changed from a contractile phenotype in primary culture to a synthetic phenotype during successive subcultures. The change in phenotype was monitored by immunofluorescent staining of actin and myosin; by the third passage, myosin was absent. The pattern of GAG synthesis at this stage, however, was still similar to the primary culture with the pericellular GAG characterized by a large amount of HS (58%) and moderate amounts of DS (26%). By the seventh and subsequent subcultures, by which stage actin exhibited a decreased staining intensity, HS was replaced by DS as the dominant pericellular GAG. Superimposed on these changes were other changes, of the type discussed earlier, associated with alterations in cell density.

Yamamoto et al.[15] also compared primary cultures to medial explants of the rat aortic wall and found that over a 48-hr incubation period the explants synthesized much less HS and more HA. Although no comment was made on the state of the SMC in the explants, presumably they would have retained the contractile phenotype since SMC of rat carotid arteries maintained under organ culture conditions retain the morphological features of the contractile phenotype for more than a week.[40] Like the aortic explants, the rat carotid in organ culture also displayed a change in GAG synthesis. Although the percentage distribution of the individual GAG was not determined, the pattern of [^{3}H] glucosamine incorporation across the vessel wall, while initially very similar to carotids labeled in vivo, changed progressively with time in culture.[41] The carotids in culture, however, still had their endothelium intact, and it is likely that these cells were responsible for some of the changes in GAG synthesis by the SMC in the vessel wall.

Mozzicato et al.[42] investigated GAG synthesis in intimal-medial explants from the aorta of Cynomolgus monkey. Although it is not clear if endothelial cells were present, over the first 5 hr in culture the percentage distribution of newly synthesized GAG was the same as for that for GAG present in the wall at the start of culturing. The relative percentages were CS 53%, DS 27%, HS 12% and HA 9%. With increasing time in culture, however, the percentage distribution pattern for the newly synthesized GAG gradually changed, and by 22 hr the amount of labeled DS (30%) was almost equal to CS (32%) with HS (22%) and HA (16%) still contributing the least to total labeled GAG.

Thus even short-term maintenance of artery wall in organ culture, where the cell/matrix relationship more closely resembles normal wall and where the SMC still display the morphological features of a contractile phenotype, leads to changes in GAG synthetic patterns. But it remains to be determined whether or not these changes occurring in whole vessel wall in organ culture are truly dependent or independent of SMC phenotype.

D. Degradation

A potential source of the variation in the amounts of individual GAG produced by SMC in culture is a difference in the rate of degradation rather than a difference in the rate of synthesis. In most studies, while it is acknowledged that labeling levels or actual amounts of GAG are the net result of synthesis and degradation, it is usually assumed that any differences in GAG levels between control and test samples are due to altered rates of synthesis. The evidence from those studies which have investigated turnover would support, in general, this premise.

First, studies with both SMC[43] and fibroblasts[44] have shown that most newly synthesized GAG passes rapidly from the intracellular pool to the pericellular pool and into the medium. Addition of exogenous GAG has shown that there is virtually no degradation of GAG in the medium. Pulse chase experiments have shown, however, that 30 to 50% of the total labeled SGAG is degraded by between 1 and 5 days and that this loss exceeds amounts initially present in the intracellular pool. Thus some degradation is occurring in the pericellular pool, and this GAG probably contributes most to the degradative pool. Diehl et al.[45] showed that inhibitors of lysosomal function had no effect on the rate of loss of material from the intracellular pool and in view of the absence of degradation in the medium, they also concluded that degradation occurs from the pericellular compartment. More recently, Volker et al.[46] have shown conclusively that cell surface proteoglycans, visualized by attached gold particles, are endocytosed via coated and noncoated cell surface pits and degraded in lysosomes.

Studies on fibroblasts[44] have further shown that the turnover of individual GAG in the pericellular layer varies. Heparan sulfate is retained for a longer period than DS and accumulates. With increasing time in culture, the proportion of HS in the medium also increases as it is released from the pericellular pool. Thus the proportions of the individual GAG in the medium will be dependent in part upon the length of time in culture and the labeling times, whereas the proportions in the pericellular layer will be additionally influenced by rates of degradation.

Although turnover studies have been made of cells maintained under constant conditions, there have been few studies attempting to determine if changed levels of GAG under experimental conditions are the result of changes in rates of synthesis or rates of degradation. Merrilees and Scott[41] have shown that the pattern of incorporation of [^{3}H] glucosamine across the wall of the rat carotid artery maintained in organ culture is markedly different in the presence and absence of endothelium. Both pulse and continuous labeling, however, showed that net rates of degradation were the same under both conditions. Rates of degradation for some of the individual GAG may have varied, but if that did occur they were offset by changed rates for others. Overall it was concluded that changed labeling levels were due to changes in rates of synthesis.

III. EFFECT OF FACTORS RELATED TO ATHEROGENESIS

The primary purpose of many of the studies on GAG production by SMC in culture has been to investigate the effect of factors related to atherogenesis in isolation from the complexity of the in vivo environment. Although some of the results are contradictory, after elimination of effects due to methodology and to culture conditions, it would appear that in general there is a good correlation between factors associated with atherogenesis in vivo and their effects on GAG production in vitro (Table 2). In vivo, developing atherosclerotic lesions are characterized by increased amounts of GAG compared with unaffected parts of vessel walls and especially an increase in SGAG relative to HA. The increase in total GAG may not be particularly large, but the change in the HA to SGAG ratio can be a valuable indicator to changed vessel structure. Thus a decrease in the HA to SGAG ratio is associated with factors that increase the risk of atherosclerosis, while the converse is generally true for factors which decrease the risk.[61]

A. Lipoproteins

The effect of lipoproteins on GAG metabolism has been the most controversial of all factor studies, probably due in large part to the fact that there have been more studies on lipoproteins than any other factor and thus more scope for variability in methodology and culture conditions. For example, the recovery of GAG by methods using cetylpyridinium chloride is reduced in the presence of hyperlipidemic serum.[47] Also, lipoproteins may effect cell growth, thus making interpretation of changes in GAG more difficult. Where cell growth has been carefully monitored, however, it is evident that lipoproteins can have a direct effect of GAG synthesis.

The most significant finding is that the atherogenic low-density lipoproteins (LDL) have an opposite effect from the antiatherogenic high-density lipoproteins (HDL). The most consistent finding is that LDL produces a decrease in the HA to SGAG ratio, while HDL produces an increase.

In the presence of LDL or in the presence of hyperlipidemic serum, the decrease in the ratio may be achieved either by a decrease in HA with little change to the amount of SGAG[48-50] or by an increase, usually small, in the amount of SGAG.[51,52] It should be noted, however, that while it is possible to generalize in this way, the effect of LDL on SMC is complex.

Wosu et al.[17] have shown that the LDL-stimulated increase in GAG is more marked when the cells are proliferating but that high levels of LDL have a cytotoxic effect, a feature noted previously for SMC[53] and for other cell types.[54,55] Similar results were obtained for both SMC and fibroblasts from adults, but fetal SMC responded differently. With fetal SMC, low levels of LDL in the presence of a low serum concentration had a stimulatory effect on SMC growth, as well as on GAG secretion; however, with

Table 2
EFFECT OF ATHEROSCLEROTIC RISK FACTORS ON GLYCOSAMINOGLYCAN PRODUCTION BY SMC IN CULTURE

		Effect on GAG levels[a]						
Risk factor	Source of SMC	HA	HS	DS	CS A/C	Total	HA/SGAG[a]	Ref.
LDL[b]	Human	—	↔	↔	↔	↔	—	56
LDL	Human							
	Adult	—	—	—	—	↑	↓	17
	Fetal	—	—	—	—	↔	—	
LDL	Bovine	—	↑	(↑)		↑	—	51
LDL-Sera	Human (fetal)	↔	(	↔	)	—	↔	59
LDL-Sera	Human (fetal)	↓	(	↔	)	—	↓	48
LDL-Sera (Type IIA hyperlipidemia)	Human (fetal)	↓	(	↔	)	—	↓	49
LDL-Sera (Type IIA hyperlipidemia)	Human (fetal)	↓	(	↔	)	—	↓	50
LDL-Sera (long-term culture, 28 days)	Rabbit	↓	(	↑	)	—	↓	52
HDL[c]	Bovine	—	↓	(↓)		↓	—	51
HDL-Sera	Human (fetal)	↓	(	↓	)	—	↑	59
HDL-Sera	Human (fetal)	↓	↓	↓	↓	↓	↑	60
HDL-Sera	Human (fetal)	↔	(	↓	)	—	↑	48
HDL-Sera	Human (fetal)							
Lumberjacks		↓	(	↓	)	—	↑	49
Runners		↔	(	↓	)	—	↑	
Hormones	Human (fetal)							
Cortisol		↓	(	↔		—	↓	48, 61
Estradiol		↔	(	↔		—	↔	
Testosterone		↔	(	↔		—	↔	
Insulin		↔	(	↔		—	↔	
Prostaglandins (E_1, E_2, $F_1\alpha$, $F_2\alpha$)	Rabbit	↑E_2	(	↑	)	—	—	70
Hypoxia	Rabbit	↑	(	↑	)		↓	72
Mechanical forces								
Cyclic stretching	Rabbit	↑	—	↔	↑	—	—	29
Centrifugation	Porcine	↑	↑	↑	↑	↑	↓	13
Endothelial cells	Porcine	↑	↑	↑	↑	↑	↑	26
Low density		↑	↑	↑	↑	↑	↑	26
High density		↔	↑	↑	↑	↑	↓	102
Ascorbic acid	Rat	—	↓	↑	↑	—	—	95
(50 μg/mℓ)	Bovine	—	↓	↑	↑	—	—	

[a] ↑ increase; ↓ decrease; ↔ no change; — not determined.
[b] Low-density lipoprotein.
[c] High-density lipoprotein.

increasing concentration of LDL, both cell number and GAG secretion were significantly reduced. Interestingly, confluent cultures of SMC did not respond to LDL with an increase in GAG secretion. Sudhakaran et al.[56] also found that LDL had no effect on confluent cultures of SMC, but Ehrlich and Murray[57] found that confluent primary cultures of SMC from rabbit and rhesus monkey did respond to LDL by increasing very significantly SGAG synthesis.

This variability in the response of SMC to LDL and to hyperlipidemic serum illustrates very clearly the complexity of the system and the necessity for further detailed studies. For example, the effect of LDL on the individual SGAG species has not been fully investigated, although Pillai et al.[51] did note that increases in HS and the other SGAG were of a similar magnitude. The effects of the various subspecies of LDL are a further avenue for future studies since it has been shown that whereas the small LDL subspecies exhibit no mitogenic action on SMC growth in culture, the large subspecies can have a marked proliferative action.[58]

The effect of HDL[51] or high-density lipoproteinemic sera[48,49,59,60] on SMC is more marked and less variable than LDL. HDL decreases the synthesis of SGAG[48,49,51,59] and may[59,60] or may not[48,49] decrease the synthesis of HA. Where HA is decreased, however, the ratio of HA to SGAG still favors HA. The effects on GAG synthesis can also be more clearly separated from cell density effects than for LDL. Tammi et al.[60] found that HDL-rich sera, collected from lumberjacks, had no effect on [^{3}H] thymidine incorporation by human SMC up to a concentration of 10%, although at 15% there was significant inhibition compared with control sera. Within the range of serum concentration where [^{3}H] thymidine incorporation was not affected, the synthesis of SGAG decreased significantly, and the individual SGAG were decreased by the same extent. HA was also decreased although at serum concentrations up to 5% the synthesis of HA was increased compared to the absence of serum. The mechanism by which HDL-sera affects GAG synthesis is not known, but there is some evidence that the effect may be dependent on the platelet-derived growth factor (PDGF).[48]

B. Hormones

Little is known about the effects of hormones on GAG synthesis. Rönnemaa et al.[49] in their studies on the effects of lipoproteins also included results on the effects of a number of hormones, including cortisol, estradiol, testosterone, and insulin. The only hormone to affect GAG synthesis was cortisol, which, at concentrations slightly above physiological level, decreased the synthesis of HA by 50% but had no effect on SGAG. Further studies showed that the decrease in HA was not due to a direct effect on the HA synthesizing system.[61] Rather, cortisol may act by changing the properties of the cell membrane,[62] the site where hyaluronate is synthesized.[63] Elevated levels of cortisol and other glucorticoids are considered to be risk factors for atherosclerosis.[64]

The role of estrogens in atherosclerosis is not clear. While epidemiological studies indicate that premenopausal women have less coronary artery disease than men, there are no data to support such a protective function. To the contrary, elevated levels seem to be associated with increased risk.[65] Interestingly, administration of estrogen or estrogen-progesterone increases the GAG content of rabbit thoracic aorta and with a marked decrease in the HA to CS ratio[66] — which implies an increased risk. The percentage distribution of the individual GAG after treatment, however, resembled that for canine thoracic aorta, which is resistant to atherosclerosis. Thus interpretation of the significance of the results is difficult. The response of vessel wall to estrogen may also be different from other tissues. It has been shown (in mice) that estradiol administration markedly increases HA content of dermal tissues while not affecting DS.[67] From the limited data available, however, it would seem that GAG production by SMC in culture is not affected by estradiol.[49]

C. Prostaglandins

The synthesis of prostaglandin E_2 is elevated in atherosclerotic rabbit aortas[68] and also in the aortas of atherosclerotic-susceptible White Carneau pigeons when compared to the resistant Show Racer breed,[69] but it is not clear if prostaglandins play a direct role in lesion development. The effects of prostaglandins E_1, E_2, $F_1\alpha$, and $F_2\alpha$ on SMC in culture have been investigated by Pietila et al.[70] Although all of these prostaglandins enhanced the incorporation of [^{3}H] glucosamine into SGAG by about 50%, they also inhibited [^{3}H] thymidine uptake; thus some of the increase may be accounted for by a lower cell density. The effect of prostaglandins on HA, however, indicates that density effects were not solely responsible since for all prostaglandins except E_2 there was less change in HA than SGAG, whereas at lower densities relatively more HA would be expected. Prostaglandin E_2 did stimulate HA, but SGAG also increased by a similar amount again indicating that the effect was probably independent of cell density.

D. Hypoxia

In experimental animals, systemic hypoxia can induce atherosclerosis with an associated increase in the synthesis and accumulation of GAG.[71] In culture the synthesis of GAG by rabbit SMC incubated in medium with oxygen saturation lowered from 20% to 2 to 3% is increased, especially SGAG (twofold increase), although HA is also increased significantly.[72] The amount of cell protein changed little with hypoxia; thus the effect on GAG levels was probably independent of cell density and may even have bee underestimated since the hypoxic cultures had a lower pH than controls. As previously discussed, pH is positively correlated with GAG synthesis.

E. Mechanical Forces

An association has been demonstrated between hypertension and increased SGAG.[73-76] While this association may not be the result of a direct effect of mechanical stress acting on the SMC in the vessel wall, connective tissues in general are very responsive to their mechanical environment.[11] In cell culture there is evidence that SMC can respond to changes in physical forces acting on the cells. Leung et al.[29] have shown that rabbit SMC grown on purified elastin membranes and subjected to repeated elongation and relaxation increase their synthesis of HA and CS6 threefold — and without any change in cell number, compared to stationary cultures. DS and CS4, however, did not increase. SMC also respond to increased pressure generated by centrifuging cultures at low speeds ($45 \times g$).[13] Confluent cultures, centrifuged for 16 hr/day for a total spin time of 48 hr over a 72-hr period, increased their GAG by 50%. All GAG increased, but there was a greater proportional increase for SGAG (60 to 90%) than for HA (-35%).

Thus, SMC are capable of responding to the mechanical environment, but interpretation of these results must be made with considerable caution. Apart from the fact that the experimental designs used in the above experiments may be quite inappropriate models for SMC subjected to stress in vivo, changes in GAG levels observed with hypertension may be mediated not through a direct effect but through, for example, the endothelium. Endothelial cells are known to respond to acute hypertension by increasing their turnover,[77] and the increase in GAG (which is primarily intimal) may be mediated by factors released by active endothelium.

F. Endothelium

Endothelial cells in culture produce a factor or factors able to stimulate GAG synthesis by SMC.[26,78] This stimulation occurs both with endothelial cells grown in coculture with SMC and with growth medium conditioned on endothelial cells and then placed on SMC. Generally the increase in HA is greater than that for SGAG; HA usually shows at least a two- to threefold increase, although the increase may be as much as seven- to eightfold over controls, while SGAG usually doubles in amount. This stimulation by endothelial cells also occurs across species[26] and also with other cells such as fibroblasts.[78] More recent evidence indicates that conditioned media from low-density cultures of endothelial cells is more effective than media from confluent cultures, although the latter media still stimulates SGAG synthesis.[102]

These results suggest that in vivo endothelium could play a role in modulating GAG synthesis by SMC in the intima or inner media and that the modulation may be dependent upon the state of the endothelium. Similar ideas have been advanced for the control of SMC growth by endothelium. In culture, endothelial cells produce both stimulatory[79] and inhibitory[80] factors for SMC growth depending upon culture conditions. There is also some support for the idea of endothelial control of SMC GAG from studies of rat carotid artery maintained in organ culture with endothelium present, absent, and regenerating.[41] In the presence of mitotically active endothelium, the synthesis of GAG in the inner layer of the media is increased compared with the rest of the media. In the absence of endothelium, removed by air-drying, there is a marked drop in the rate of synthesis in the inner layer, but this returns to normal following regeneration of endothelium. Throughout these changed levels of GAG the rates of degradation, as measured by pulse and continuous labeling, remained the same as for controls. Furthermore, throughout the various experimental conditions, there was no significant growth of SMC. Similar changes in intimal medial GAG levels in the absence and presence of regenerating endothelium have been observed in experimental animal models,[81,82] but the changes observed in organ culture occurred in serum-free medium and thus independently of factors such as PDGF. Endothelium in an active state may thus play a direct role in modulating GAG levels and independently from SMC growth.

G. Platelets and Monocytes

Platelets are thought to play a pivotal role in the initiation of SMC growth and lesion development through the release of PDGF. Platelets also produce connective tissue activating peptides (CTAP III and CTAP-P2)[83,84] that stimulate GAG synthesis, especially HA, in fibroblasts. The effect on SMC is not known, but it is possible that they would respond in a similar manner.

The role of monocytes in atherogenesis, especially the early stages, is currently receiving increased attention.[85-87] While their primary role appears to be one of lipid uptake, in culture they have been shown to release a growth factor for SMC;[88] therefore, in addition to forming foam cells they may contribute to lesion development by stimulating SMC growth. Their effect on GAG synthesis, however, is not known although it is possible that they could play a role in degradation since peripheral blood mononuclear cells can release distinct latent enzymes capable of degrading both collagen and proteoglycans.[89]

H. Ascorbic Acid

Ascorbic acid (AA) deficiency is considered to be a risk factor for atherosclerosis.[90,91] It must be noted, however, that overt AA deficiency, or scurvy, is a rare occurrence, which implies that if AA does play a role in the pathogenesis of atherosclerosis, then either there are local deficiencies in vessel walls or AA affects vessel wall structure in prescorbutic states. With true AA deficiency it is clearly apparent that connective tissue structure and organization is affected, and it is well known that AA is required for hydroxylation of collagen, without which proper fibril formation will not take place.[92]

AA also affects GAG levels in vessel walls. For example Verlangieri and Stevens[93] showed that aortic GAG levels (and especially CS4) in rabbits fed an atherogenic diet supplemented with AA were increased compared with animals on the same diet but without AA. The increased GAG levels here were interpreted as evidence in support of the idea that AA is necessary to maintain adequate levels of GAG in vessels to maintain the structural integrity and resist atherosclerotic changes. The lower levels of GAG found in animals on the atherogenic diet without AA were thus considered to contribute to the process of atherogenesis.

Evidence from cell culture studies supports the idea that AA plays a role in maintaining the production of extracellular materials by SMC. Like fibroblasts, SMC in culture require adequate levels of AA for the formation of insoluble collagen.[94] Furthermore, addition of AA to SMC cultures also enhances both the synthesis and degree of sulfation of SGAG, except for HS, and independently of increased collagen synthesis.[95] In these studies morphology and rates of proliferation were not affected by the addition of AA, although others[96] have found that AA affects both parameters of bovine SMC in culture, with cell doubling time being decreased from 62 to 47 hr in the presence of AA and cells changing from an elongated to a polygonal shape. Further, in this latter study there was evidence that HS increased.

In the above studies, AA was added to a level of 50 μg/mℓ. In some unpublished studies using porcine and rat SMC, increasing concentrations of AA ranging from 25 to 300 μg/mℓ were found to have a biphasic effect on GAG synthesis.[16] At concentrations of 25 μg/mℓ and 50 μg/mℓ all GAG, including HA, were stimulated to a small degree (25 to 30%), but above 50 μg/mℓ GAG synthesis was decreased markedly. Cell density was also reduced but only at 200 μg/mℓ and 300 μg/mℓ, whereas the reduction in GAG was apparent at 100 μg/mℓ where cell densities and morphologies were the same as for lower concentrations of AA. A concentration of 50 μg/mℓ, however, is above the normal physiological level of 0.7 to 1.2 mg/100 mℓ of plasma. It would appear, therefore, that GAG synthesis may be increased by AA supplementation even if only to a small degree.

These studies on AA also illustrate the importance of being able to compare changes in GAG levels in culture with changes in GAG levels in vivo. Otherwise the increased GAG levels in culture could be interpreted as increasing the risk of atherosclerosis by increasing available GAG for binding to lipoproteins rather than bringing GAG levels up to a level adequate to maintain normal vessel structure. Interestingly, while studies on changes in collagen synthesis have been invariably conducted with AA-supplemented media, few studies on GAG metabolism have included AA in the media. Stimulation of SMC GAG synthesis by endothelial cells, however, still occurs in media supplemented with AA.[78]

IV. PROTEOGLYCAN SYNTHESIS

Only recently have attempts been made to characterize in detail the types of PG synthesized by SMC in culture. Studies on SMC from bovine,[97] porcine,[9] and primate[10] aortas have shown that at least two distinct size classes of proteoglycans are produced with physicochemical properties similar to PG isolated from intact vessel walls. More recently, Arumugham and Anderson[98] have established for bovine SMC in culture that although both the medium and cell layer contain predominately CS proteoglycans, the cell layer contains a higher proportion of HS and DS proteoglycans. As yet, however, there is virtually no information on the effects of either culture conditions or atherogenic risk factors on PG production. Nevertheless, the demonstration that SMC in culture secrete PG with characteristics similar to those present in vivo is important since investigations on factors suspected of modulating PG synthesis can now be carried out using the advantages of cell culture and with some confidence that the cells will retain their differentiated function.

V. CONCLUDING REMARKS

It is apparent from the results of these cell culture studies that many of the changes to the amounts and proportions of GAG seen in intact vessel walls can be explained by the direct effects of stimulatory or inhibitory factors on GAG synthesis by the SMC. It is also clear that the synthesis of GAG by SMC in culture is highly variable and that before concluding that any particular factor has a direct effect, care must be taken to ensure that differences between control and test cultures are not due to changes in culture conditions, especially changes in cell density, serum concentration, and pH.

This review also highlights a number of apparent gaps in our knowledge of SMC synthesis by SMC in culture. Perhaps foremost is that virtually all studies have been carried out using SMC from a single vessel, the aorta, and while it is likely that SMC from other vessels are no different from aortic SMC, it would seem worthwhile to examine SMC from other sources such as coronary arteries.

It is also notable that while SMC from fetal human aortas have been used fairly extensively for the investigation of factors associated with atherosclerosis, there is little information available on GAG production by adult human SMC. Furthermore, in only one study on human SMC[60] have the individual SGAG been analyzed separately. Of further interest is the finding that human SMC in culture produce a large amount of HA, much more than would be expected from amounts present in the whole aorta.[31,99] SMC from nonhuman primates and from many other animals are characterized by a low proportion of HA; thus it would be of interest to know if HA production by human SMC in culture is a specific trait or relates to cell density. Although many of the cultures with a high HA production were apparently confluent, the effect of cell density on HA production for these SMC is not known.

Although it has been shown that SMC of a contractile phenotype produce a slightly different profile of SMC compared to synthetic SMC,[15] further studies are required to confirm and extend these findings. In particular, a strong argument can be advanced for investigating the effect of atherosclerotic risk factors on contractile SMC. In the vast majority of cell culture studies, the SMC have been of the synthetic phenotype, but Chamley-Campbell et al.[101] have shown that contractile SMC do not respond to PDGF, hyperlipidemic LDL, and other serum-derived mitogens. Once they have modulated to a synthetic phenotype, however, they are immediately responsive and undergo proliferation. It would be of interest to determine if changes in GAG production by SMC are similarly dependent on phenotype.

Further studies are also warranted on the effects of other factors such as hormones and ascorbic acid, and it may also be appropriate with these and other studies to make more extensive use of vessels in organ culture where the cell/matrix relationship more closely resembles the condition in vivo. Finally, it is important to reiterate that a change in the HA to SGAG ratio is just as important as a change in GAG amounts and that whenever possible the effect of factors on both HA and SGAG should be determined.

REFERENCES

1. Kramsch, D. M., Biochemical changes in the arterial wall in atherosclerosis with special reference to connective tissue: promising experimental avenues for their prevention, in *Connective Tissues in Arterial and Pulmonary Disease*, McDonald, T. F. and Chandler, A. B., Eds., Springer-Verlag, New York, 1981, 95.
2. Pietilä, K. and Nikkari, T., Role of the arterial smooth muscle cell in the pathogenesis of atherosclerosis, *Med. Biol.*, 61, 31, 1983.
3. Amenta, J. S. and Waters, L. L., The precipitation of serum lipoproteins by mucopolysaccharides extracted from aortic tissue, *Yale J. Biol. Med.*, 33, 112, 1960.
4. Gerö, S., Gergely, J., Devenyi, T., Jakob, L., Szekely, J., and Virag, S., Role of intimal mucoid substances in the pathogenesis of atherosclerosis. I. Complex formation in vitro between mucopolysaccharides from atherosclerotic aortic intimas and plasma β-lipoprotein and fibrinogen, *J. Atheroscler. Res.*, 1, 67, 1961.
5. Srinivasan, S. R., Lopez, S. A., Radhakrishnamurthy, B., and Berenson, G. S., complexing of serum pre-β- and β-lipoproteins and acid mucopolysaccharides, *Atherosclerosis*, 12, 321, 1970.
6. Iverius, P. H., The interaction between human plasma lipoproteins and connective tissue glycosaminoglycans, *J. Biol. Chem.*, 247, 2607, 1972.
7. Iverius, P. H., Possible role of glycosaminoglycans in the genesis of atherosclerosis, in *Atherogenesis: Initiating Factors*, Ciba Symposium No. 12, Elsevier, Amsterdam, 1973, 185.
8. McMurtrey, J., Radhakrishnamurthy, B., Dalferes, E. R., Berenson, G. S., and Gregory, J. D., Isolation of proteoglycan-hyaluronate complexes from bovine aorta, *J. Biol. Chem.*, 254, 1621, 1979.
9. Horn, M. C., Breton, M., Deudon, E., Berrou, E., and Picard, J., The structural characterisation of proteoglycans of cultured aortic smooth muscle cells and arterial wall of the pig, *Biochim. Biophys. Acta*, 755, 95, 1983.
10. Wight, T. N. and Hascall, V. C., Proteoglycans in primate arteries. III. Characterization of the proteoglycans synthesized by arterial smooth muscle cells in culture, *J. Cell Biol.*, 96, 167, 1983.
11. Flint, M. H., Gillard, G. C., and Merrilees, M. J., The effect of local physical environmental factors on connective tissue organization and glycosaminoglycan synthesis, in *Fibrous Proteins: Scientific, Industrial and Medical Aspects*, Vol. 2, Parry, D. A. D. and Creamer, L. K., Eds., Academic Press, London, 1980, 107.
12. Pietilä, K., Incorporation of [^{3}H] glucosamine into glycosaminoglycans in different cell culture conditions by rabbit aortic smooth muscle cells, *Biochim. Biophys. Acta*, 677, 318, 1981.
13. Merrilees, M. J., Merrilees, M. A., Birnbaum, P. S., Scott, P. J., and Flint, M. H., The effect of centrifugal force on glycosaminoglycan production by aortic smooth muscle cells in culture, *Atherosclerosis*, 27, 259, 1977.
14. Wight, T. N. and Ross, R., Proteoglycans in primate arteries. II. Synthesis and secretion of glycosaminoglycans by arterial smooth muscle cells in culture, *J. Cell Biol.*, 67, 675, 1975.
15. Yamamoto, H., Kanaide, H., and Nakamura, M., Metabolism of glycosaminoglycans of cultured rat aortic smooth muscle cells altered during subculture, *Br. J. Exp. Pathol.*, 64, 156, 1983.
16. Merrilees, M. J. and Scott, L., unpublished data, 1984.
17. Wosu, L., Parisella, R., and Kalant, N., Effect of low density lipoprotein on glycosaminoglycan secretion by cultured human smooth muscle cells and fibroblasts, *Atherosclerosis*, 48, 205, 1983.
18. Oakes, B. W., Batty, A. C., Handley, C. J., and Sandberg, L. B., The synthesis of elastin, collagen, and glycosaminoglycans by high density primary cultures of neonatal rat aortic smooth muscle. An ultrastructural and biochemical study, *Eur. J. Cell Biol.*, 27, 34, 1982.
19. Tomida, M., Koyama, H., and Ono, T., A serum factor capable of stimulating hyaluronic acid synthesis in cultured rat fibroblasts, *J. Cell. Physiol.*, 91, 323, 1977.
20. Gamse, G., Fromme, H. G., and Kresse, H., Metabolism of sulphated glycosaminoglycans in cultured endothelial cells and smooth muscle cells from bovine aorta, *Biochim. Biophys. Acta*, 544, 514, 1978.
21. Pietilä, K., Ylä-Herttuala, S., Jaakkola, O., and Nikkari, T., Metabolism of glycosaminoglycans and lipids in smooth muscle cells from atherosclerotic rabbit aortas in culture, *Atherosclerosis*, 37, 449, 1980.
22. Pietilä, K., Ylä-Herttuala, S., Rantala, I., and Nikkari, T., Characterisation of cells cultured from advanced atherosclerotic lesions of the rabbit, *Med. Biol.*, 60, 221, 1982.
23. Wight, T. N., Differences in the synthesis and secretion of sulphated glycosaminoglycans by aorta explant monolayers cultured from atherosclerotic-susceptible and resistant pigeons, *Am. J. Pathol.*, 101, 127, 1980.
24. Schmidt, A., Grünwald, J., and Buddecke, E., [^{35}S] Proteoglycan metabolism of arterial smooth muscle cells cultured from normotensive and hypertensive rats, *Atherosclerosis*, 45, 299, 1982.

25. Taniguchi, N., Isolation and analysis of glycosaminoglycans, in *Glycosaminoglycans and Proteoglycans in Physiological Processes of Body Systems*, Varma, R. S. and Varma, R., Eds., S. Karger, Basel, 1982.
26. Merrilees, M. J. and Scott, L., Interaction of aortic endothelial and smooth muscle cells in culture, *Atherosclerosis*, 39, 147, 1981.
27. Engel, U. R., Glycosaminoglycans in the aorta of six animal species, *Atherosclerosis*, 13, 45, 1971.
28. Stefanovich, V. and Akiyama, K., Comparative studies of aortic acid mucopolysaccharides in fifteen species, *Comp. Biochem. Physiol.*, 34, 125, 1970.
29. Leung, D. Y. M., Glagov, S., and Mathews, M. B., Cyclic stretching stimulates synthesis of matrix components by arterial smooth muscle cells in vitro, *Science*, 191, 475, 1976.
30. Namiki, O., Faris, B., Tschopp, F., Fuglistaller, P., Hollander, W., Franzblau, C., and Schmid, K., Synthesis of glycosaminoglycans by cultured rabbit smooth muscle, *Biochemistry*, 19, 1900, 1980.
31. Gardais, A., Picard, J., and Hermelin, B., Glycosaminoglycan (GAG) distribution in aortic wall from five species, *Comp. Biochem. Physiol. B*, 44, 507, 1973.
32. Curwen, K. D. and Smith, S. S., Aortic glycosaminoglycans in atherosclerosis-susceptible and resistant pigeons, *Exp. Mol. Pathol.*, 27, 121, 1977.
33. Wight, T. N. and Ross, R., Proteoglycans in primate arteries. I. Ultra-structural localization and distribution in the intima, *J. Cell Biol.*, 67, 660, 1975.
34. Hassell, J. R., Robey, P. G., Barrach, H.-J., Wilczek, J., Rennard, S. I., and Martin, G. R., Isolation of a heparan sulphate-containing proteoglycan from basement membrane, *Proc. Natl. Acad. Sci. U.S.A.*, 77, 4494, 1980.
35. Heickendorff, L. and Ledet, T., Glycosaminoglycans of arterial basement membrane-like material from cultured rabbit aortic myomedial cells, *Biochim. Biophys. Acta*, 798, 276, 1984.
36. Nilsson, J., Ksiazek, T., Thyberg, J., and Wasteson, A., Cell surface components and growth regulation in cultivated arterial smooth muscle cells, *J. Cell Sci.*, 64, 107, 1983.
37. Kraemer, P. M., Heparan sulphates of cultured cells. I. Membrane associated and cell-sap species in Chinese hampster cells, *Biochemistry*, 10, 1437, 1971.
38. Kraemer, P. M., Heparan sulphates of cultured cells. II. Acid-soluble and precipitable species of different cell lines, *Biochemistry*, 10, 1445, 1971.
39. Kraemer, P. M., Mucopolysaccharides: cell biology and malignancy, in *Surfaces of Normal and Malignant Cells*, Hynes, R. O., Ed., John Wiley & Sons, New York, 1979, 149.
40. Merrilees, M. J. and Scott, L., Organ culture of rat carotid artery: maintenance of morphological characteristics and of pattern of matrix synthesis, *In Vitro*, 18, 900, 1982.
41. Merrilees, M. J. and Scott, L., Effects of endothelial removal and regeneration on smooth muscle glycosaminoglycan synthesis and growth in rat carotid artery in organ culture, *Lab. Invest.*, 52, 409, 1985.
42. Mozzicato, P., Faris, B., Hollander, W., Franzblau, C., and Schmid, K., A method to evaluate the biosynthesis of glycosaminoglycans by the aorta of Cynomolgus Monkey, *Atherosclerosis*, 45, 359, 1982.
43. Deudon, E., Breton, M., Berrou, E., and Picard, J., Metabolism of glycosaminoglycans in cultured smooth muscle cells from pig aorta, *Biochimie*, 62, 811, 1980.
44. Vogel, K. G. and Kendall, V. F., Cell-surface glycosaminoglycans: turnover in cultured human embryo fibroblasts (IMR-90), *J. Cell. Physiol.*, 103, 475, 1980.
45. Diehl, T., Scott-Burden, T., and Gevers, W., Sulphate turnover of surface proteoglycans in cultured rat smooth muscle cells, *Biochim. Int.*, 6, 29, 1983.
46. Volker, W., Schmidt, A., Robenek, H., and Buddecke, E., Binding and degradation of proteoglycans by cultured arterial smooth muscle cells. I. Endocytosis and intracellular translocation of proteoglycan-gold conjugates, *Eur. J. Cell Biol.*, 34, 110, 1984.
47. Pietilä, K. and Nikkari, T., Hyperlipidemic rabbit serum reduces recovery of acidic glycosaminoglycans from tissue culture medium, *Anal. Biochem.*, 116, 317, 1981.
48. Rönnemaa, T., Serum lipoprotein composition, platelet factor and arterial smooth muscle cells, *Acta Med. Scand. Suppl.*, 642, 55, 1980.
49. Rönnemaa, T., Järvelainen, H., Lehtonen, A., Tammi, M., Larjava, H., Saarni, H., Vihersaari, T., and Viikari, J., Serum lipoprotein composition, hormones, and the synthesis of glycosaminoglycans by human aortic smooth muscle cells, *Artery*, 8, 323, 1980.
50. Järveläinen, H., Rönnemaa, T., Tammi, M., Vihersaari, T., Lehtonen, A., and Viikari, J., Type IIA Hyperlipoproteinemic sera decreases the synthesis of hyaluronic acid by cultured human aortic smooth muscle cells, *Atherosclerosis*, 39, 61, 1981.
51. Pillai, S., Donnelly, P. V., Mourao, P. A. S., Di Ferrante, N., and Eskin, S. G., Effect of high density lipoproteins and low density lipoproteins on the glycosaminoglycans metabolism of cultured bovine smooth muscle cells, *Biochem. Int.*, 1, 55, 1980.

52. Pietilä, K., Long-term effect of hyperlipidemic serum on the synthesis of glycosaminoglycans and on the rate of growth of rabbit aortic smooth muscle cells in culture, *Atherosclerosis,* 42, 67, 1982.
53. Hessler, J. R., Robertson, A. L., and Chisolm, G. M., LDL induced cytotoxicity and its inhibition by HDL in human vascular smooth muscle and endothelial cells in culture, *Atherosclerosis,* 32, 213, 1979.
54. Curtiss, L. K. and Edginton, T. S., Regulatory serum lipoproteins — regulation of lymphocyte stimulation by a species of low density lipoprotein, *J. Immunol.,* 116, 1452, 1976.
55. Henricksen, T., Evensen, S. H., and Carlander, B., Injury to human endothelial cells in culture induced by low density lipoprotein, *Scand. J. Clin. Lab. Invest.,* 39, 361, 1979.
56. Sudhakaran, P. R., Prinz, R., Filipovic, I., von Figura, K., and Buddecke, E., Homologous low density lipoprotein does not affect proteoglycan metabolism of cultured skin fibroblasts and arterial smooth muscle cells, *Hoppe Seyler's Z. Physiol. Chem.,* 361, 129, 1980.
57. Ehrlich, K. and Murray, M., Effect of low density lipoprotein on proteoglycan synthesis by aorta cells in culture, *Experientia,* 34, 179, 1978.
58. Fless, G. M., Kirchhausen, T., Fischer-Dzoga, K., Wissler, W., and Scanu, A. M., Serum low density lipoproteins with mitogenic effect on cultured aortic smooth muscle cells, *Atherosclerosis,* 41, 171, 1982.
59. Tammi, M., Rönnemaa, T., Vihersaari, T., Saarni, H., Lehtonen, A., and Viikari, J., Effect of sera from hyperlipidemic subjects and high-density-lipoproteinemic runners on the synthesis of DNA and glycosaminoglycans by cultured human aortic smooth muscle cells, *Med. Biol.,* 57, 118, 1979.
60. Tammi, M., Rönnemaa, T., Vihersaari, T., Lehtonen, A., and Viikari, J., High density lipoproteinemia due to vigorous physical work inhibits the incorporation of [^{3}H] thymidine and the synthesis of glycosaminoglycans by human aortic smooth muscle in culture, *Atherosclerosis,* 32, 23, 1979.
61. Larjava, H., Saarni, H., Tammi, M., Penttinen, R., and Ronnemaa, T., Cortisol decreases the synthesis of hyaluronic acid by human aortic smooth muscle cells in culture, *Atherosclerosis,* 35, 135, 1980.
62. De Duwe, C., Walleaux, R., and Wibo, M., Effect of fat-soluble compounds on lysosomes in vitro, *Biochem. Pharmacol.,* 9, 97, 1972.
63. Prehm, P., Hyaluronate is synthesized at plasma membranes, *Biochem. J.,* 220, 597, 1984.
64. Troxler, R., Sprague, E. A., Albanese, R. A., Fuchs, R., and Thompson, A. J., The association of elevated plasma cortisol and early atherosclerosis as demonstrated by coronary angiography, *Atherosclerosis,* 26, 151, 1977.
65. Phillips, G. B., Sex hormones, risk factors and cardiovascular disease, *Am. J. Med.,* 65, 7, 1978.
66. Aikawa, J., Munakata, H., Isemura, M., Ototani, N., and Yosizawa, Z., Hormonal effects on glycosaminoglycans in thoracic aortas of rabbits, *Tohoku J. Exp. Med.,* 143, 113, 1984.
67. Uzuka, M., Nakajima, K., Ohta, S., and Mori, Y., The mechanism of estrogen-induced increase in hyaluronic acid biosynthesis with special reference to estrogen receptor in the mouse skin, *Biochim. Biophys. Acta,* 627, 199, 1980.
68. Berberian, P. A., Ziboh, V. A., and Hsia, S. L., Prostaglandin E_2 biosynthesis: changes in rabbit aorta and skin during experimental atherogenesis, *J. Lipid Res.,* 17, 46, 1976.
69. Subbiah, M. T. R., Prostaglandin E biosynthesis and effect in pigeon aorta — possible role in atherogenesis, *Atherosclerosis,* 29, 487, 1978.
70. Pietilä, K., Moilanen, T., and Nikkari, T., Prostaglandins enhance the synthesis of glycosaminoglycans and inhibit the growth of rabbit aortic smooth muscle cells in culture, *Artery,* 7, 509, 1980.
71. Helin, G., Helin, P., and Lorenzen, I., The aortic glycosaminoglycans in arteriosclerosis induced by systemic hypoxia, *Atherosclerosis,* 12, 235, 1970.
72. Pietilä, K. and Jaakkola, O., Effect of hypoxia on the synthesis of glycosaminoglycans and collagen by rabbit aortic smooth muscle cells in culture, *Atherosclerosis,* 50, 183, 1984.
73. Crane, W. A. J., Sites of mucopolysaccharide synthesis in the lesions of experimental hypertension in rats, *J. Pathol. Bacteriol.,* 83, 183, 1962.
74. Hollander, W., Kramsch, D. M., Farmelant, N., and Madoff, I. M., Arterial wall metabolism in experimental hypertension of coarctation of the aorta of short duration, *J. Clin. Invest.,* 47, 1221, 1968.
75. Sharpe, D. N., Scott, P. J., and Flint, M. H., Connective tissue changes in artery walls in experimental hypertension, *Aust. N. Z. J. Surg.,* 44, 427, 1974.
76. Hauss, W. H., Tissue alterations due to experimental arteriosclerosis, in *Connective Tissue and Ageing,* Vol. 1, Vogel, H. G., Ed., Excerpta Medica, Amsterdam, 1973, 23.
77. Chobanian, A. V., The influence of hypertension and other hemodynamic factors in atherogenesis, *Prog. Cardiovasc. Dis.,* 26, 177, 1983.
78. Merrilees, M. J., Sodek, J., and Aubin, J. E., Effect of cells of epithelial rests of Malassez and endothelial cells on synthesis of glycosaminoglycans by periodontal ligament fibroblasts in vitro, *Dev. Biol.,* 97, 146, 1983.

79. Gajdusek, C., DiCorleto, P., Ross, R., and Schwartz, S. M., An endothelial cell-derived growth factor, *J. Cell Biol.*, 85, 467, 1980.
80. Castellot, J. J., Addonizio, M. L., Rosenberg, R., and Karnovsky, M. J., Cultured endothelial cells produce a heparinlike inhibitor of smooth muscle cell growth, *J. Cell Biol.*, 90, 372, 1981.
81. Richardson, M., Ihnatowycz, I., and Moore, S., Glycosaminoglycan distribution in rabbit aortic wall following balloon catheter deendothelialization. An ultrastructural study, *Lab Invest.*, 43, 509, 1980.
82. Wight, T. N., Curwen, K. D., Litrenta, M. M., Alonso, D. R., and Minick, C. R., Effect of endothelium on glycosaminoglycan accumulation in injured rabbit aorta, *Am. J. Pathol.*, 113, 156, 1983.
83. Myers, S. L. and Castor, C. W., Connective tissue activation. XV. Stimulation of glycosaminoglycan and DNA synthesis by a polymorphonuclear leukocyte factor, *Arthritis Rheum.*, 23, 556, 1980.
84. Castor, C. W. and Cobel-Geard, S. R., Connective tissue activation: evidence for a second human platelet growth factor, *Clin. Res.*, 28, 139a, 1980.
85. Gerrity, R. G., The role of the monocyte in atherogenesis. I. Transition of blood-borne monocytes into foam cells in fatty lesions, *Am. J. Pathol.*, 103, 181, 1981.
86. Gerrity, R. G., The role of the monocyte in atherogenesis. II. Migration of foam cells from atherosclerotic lesions, *Am. J. Pathol.*, 103, 191, 1981.
87. Joris, I., Zand, T., Nunnari, J. J., Krolikowski, F. J., and Majno, G., Studies on the pathogenesis of atherosclerosis. I. Adhesion of mononuclear cells in the aorta of hypercholesterolemic rats, *Am. J. Pathol.*, 113, 341, 1983.
88. Ziats, N. P. and Robertson, A. L., Effects of peripheral blood monocytes on human vascular cell proliferation, *Atherosclerosis*, 38, 401, 1981.
89. Golds, E. E., Santer, V., Killackey, J., and Roughley, P. J., Mononuclear cell factors stimulate the concomitant secretion of distinct latent proteoglycan, gelatin and collagen degrading enzymes from human skin fibroblasts and synovial cells, *J. Rheumatol.*, 10, 861, 1983.
90. Shaffer, C. F., Ascorbic acid and atherosclerosis, *Am. J. Clin. Nutr.*, 23, 27, 1970.
91. Krumdieck, C. and Butterworth, C. E., Ascorbate-cholesterol-lecithin interactions: factors of potential importance in the pathogenesis of atherosclerosis, *Am. J. Clin. Nutr.*, 27, 866, 1974.
92. Prockop, J. J., Kivirikko, K. I., Tuderman, L., and Guzman, N. A., The biosynthesis of collagen and its disorders, *N. Engl. J. Med.*, 301, 13, 1979.
93. Verlangieri, A. J. and Stevens, J. W., L-ascorbic acids: effects on aortic glycosaminoglycan ^{35}S incorporation in rabbit-induced atherosclerosis, *Blood Vessels*, 16, 177, 1979.
94. De Clerck, Y. A. and Jones, P. A., The effect of ascorbic acid on the nature and production of collagen and elastin by rat smooth-muscle cells, *Biochem. J.*, 186, 217, 1980.
95. Scott-Burden, T., Murray, E., Diehl, T., and Gevers, W., Glycosaminoglycan synthesis by smooth muscle cells cultured in the absence and presence of ascorbic acid, *Hoppe Seyler's Z. Physiol. Chem.*, 364, 61, 1983.
96. Schwartz, E., Bienkowski, R. S., Coltoff-Schiller, B., Goldfisher, S., and Blumenfeld, O. O., Changes in the components of extracellular matrix and in growth properties of cultured aortic smooth muscle cells upon ascorbate feeding, *J. Cell Biol.*, 92, 462, 1982.
97. Kresse, H., von Figura, K., Buddecke, E., and Fromme, H. G., Metabolism of sulphated glycosaminoglycans in cultivated bovine arterial cells. I. Characterization of different pools of sulphated glycosaminoglycans, *Hoppe Seyler's Z. Physiol. Chem.*, 356, 929, 1975.
98. Arumugham, R. and Anderson, J. C., Distribution in caesium chloride gradients of proteoglycans secreted by bovine aorta smooth muscle cells and study of their aggregation with hyaluronate and cartilage proteoglycan, *Biochim. Biophys. Acta*, 797, 128, 1984.
99. Aikawa, J., Munakata, H., Isemura, M., Ototani, N. and Yosizawa, Z., Comparison of glycosaminoglycans from thoracic aortas of several mammals, *Tohoku J. Exp. Med.*, 143, 107, 1984.
100. Kresse, H. and Buddecke, E., Stoffwechselheterogenität ^{14}C- und ^{35}s-markieter glycosaminoglykane (saurer mucopolysaccharide) bei in-vitro-inkubation, *Hoppe Seyler's Z. Physiol. Chem.*, 351, 151, 1970.
101. Chamley-Campbell, J. H., Campbell, G. R., and Ross, R., Phenotype-dependent response of cultured aortic smooth muscle to serum mitogens, *J. Cell Biol.*, 89, 378, 1981.
102. Scott, L. J. and Merrilees, M. J., Stimulation of smooth muscle cell glycosaminoglycan synthesis by cultured endothelial cells is dependent on endothelial cell density, *Atherosclerosis*, 1986, in press.

Chapter 10

GLYCOPROTEINS ASSOCIATED WITH BOVINE AORTIC SMOOTH MUSCLE CELLS IN CULTURE — SELECTED ASPECTS

James R.A. Leushner and M. Daria Haust

TABLE OF CONTENTS

I. INTRODUCTION

The significance of smooth muscle cells (SMC) in atherosclerosis has been well documented.[1,2] Many of the advances in our understanding of their role, both in the normal and diseased arteries, have been obtained from the in vitro tissue culture system. Central to their involvement in the disease are the components they secrete into the extracellular compartment. These consist primarily of the fibrous elements of the interstitial collagens and the elastic tissue, all of which have been characterized in detail both morphologically[3-5] and biochemically.[6-11]

Other components such as glycoproteins, glycosaminoglycans, and other substances that do not form definable fibrous elements are also constituents of the extracellular matrix produced by SMC.[6,12] Generally, SMC are "wrapped" by a morphologically definable basement membrane (BM) whose major structural constituent is type IV collagen.[13,14] Closely associated with this BM are a number of high-molecular-weight structural glycoproteins such as fibronectin[15,16] and laminin.[17,18] Several other structural glycoproteins[19] have been demonstrated in the extracellular matrix. For the purpose of this presentation, these macromolecules are defined as oligosaccharide-containing matrix proteins. Therefore, this class would include the above structural glycoproteins as well as the procollagens[20] and some nonfibrous collagens such as type IV collagen.[21]

Recent studies suggest that in addition to the interstitial and BM collagens a separate class, termed "intermediate collagens",[22] also is present in the extracellular compartment. The major member of this intermediate class is the type V collagen,[7,23] which is distributed in an "intermediate zone" between the BM-cell interface (i.e., the pericellular matrix) and the interstitial matrix. Some other glycoproteins such as fibronectin[15,16] were shown to have a similar pericellular distribution. Macromolecules with this type of distribution have been implicated in the structural organization of both the BM and the intercellular matrix, serving as an organizational lattice upon which the other components of the matrix are deposited.[19] Many of these glycoproteins are intimately associated with the cytoskeletal elements and thus may provide a structural and functional link between the cell and the extracellular matrix. In this capacity the macromolecules are important in cell migration, spreading, and differentiation.[24]

In this presentation some of our studies using bovine aortic SMC in culture will be summarized in the following areas: (1) the synthesis of oligosaccharide (oligo)-containing glycoproteins in general and the interstitial and type V procollagens in particular, (2) the binding of type V collagen to the SMC surface and the role of this collagen in the transference of forces in SMC function, and (3) biochemical and morphological comparison of the features of a polypeptide complex synthesized by SMC, which has characteristics similar to both, the type VI collagen, and the microfibrils.

II. GLYCOPROTEINS IN THE SMC MATRIX

The cellular functions of SMC in culture have been shown to be dependent on the extracellular matrix adjacent to the cell (also termed the "pericellular matrix"[25]). Regulatory roles have been proposed for the matrix produced by the SMC themselves in culture[26] and that derived from other sources.[27] Studies such as these underscore the fact that rather than playing just a simple structural role, the components of the matrix are directly involved in modifying the behavior of cells.

The oligo groups of the glycoproteins in the matrix are thought to mediate the cell-matrix interconnections,[28,29] as well as cell to cell interactions.[30] Changes in the oligos at the surface of fibroblasts have been observed during cell transformation[31] and cell growth in culture.[32-34] Modifications of the oligos on the cell surface have been shown to alter the way the cell adheres to the intercellular matrix[35] and binds certain substances.[36]

Whereas the effects of various components within the matrix on SMC growth have been documented, no information on the oligo composition of the matrix had been available. Therefore, efforts in our laboratories have been directed towards testing whether the changes of the oligos associated with glycoproteins of the matrix of SMC occur in different phases of growth in culture.[37]

A. Methodology

SMC were obtained from medial explants of bovine aorta. They were grown in M199 media supplemented with 20% fetal calf serum and an antimycotic-antibiotic mixture. The proliferative stage of SMC growth was characterized by rapidly dividing cells, as shown in Figure 1A. This phase was followed by a confluent phase with multilayered growth of "hill and valley" appearance, characteristic of SMC (Figure 1B). The latter phase of growth was characterized by uptake of [^{3}H] thymidine, as shown in Table 1, column A; the labeling of these cultures was similar to the quiescent or confluent phase observed in fibroblast cultures. During the same time interval, proliferating cells incorporated more thymidine label than those in confluent cultures.

B. Glycoproteins and Oligosaccharides

The results of a 24-hr labeling of SMC in both phases of growth with either [^{3}H] Pro or [^{3}H] Man are summarized in Table 1. The cell cultures were divided into three compartments: the medium, the detergent soluble cell layer, and the insoluble matrix remaining attached to the culture flask. As indicated in column B, the level of [^{3}H] Pro incorporation was similar in all three compartments from both types of cultures. The amount of [^{3}H] Man (Table 1, column C) was consistently higher in all three compartments from proliferating than those of confluent cultures; the largest difference was present in the soluble cell fractions.

The nature of the oligos attached to the matrix secreted by both phases of cell growth were determined by several biochemical techniques.[37] The addition of the antibiotic tunicamycin, which specifically interferes with oligosynthesis,[38] resulted in a marked decrease in Man incorporation, and the level of inhibition was greater in proliferating cultures. Glycoprotein preparations were analyzed by affinity chromatography on concanavalin A-Sepharose columns which "recognize" oligos with Man residues on the nonreducing termini or substituted in their carbon 2 positions.[39,40] The results indicated that greater amounts of label from proliferating cultures were retained by the columns than from confluent cells.

Endoglycosidases (endo) were also used to characterize the mannose-labeled oligos in matrix glycoproteins. Endo H and endo D remove simple and complex oligos, respectively.[41-43] The oligomannosyl cores generated by endo treatment of glycopeptides were analyzed by molecular seiving, and the size differences are summarized in Table 2. The oligomannosyl cores derived from endo H digestion of glycopeptides from proliferating cultures consisted of eight to ten carbohydrates; those from confluent cultures were smaller and averaged four to six carbohydrates. The endo D-derived oligomannosyl cores from preparations obtained from proliferating cultures contained primarily small-core, four to six carbohydrates and small amounts of large-core subunits. On the other hand, endo D-derived preparations from confluent cultures contained larger amounts of large core subunits (15%) than the smaller cores (10%). Similar amounts of glycopeptides remained in the high-molecular-weight region after both endo treatments. In general, therefore, the matrix from proliferating cultures contains

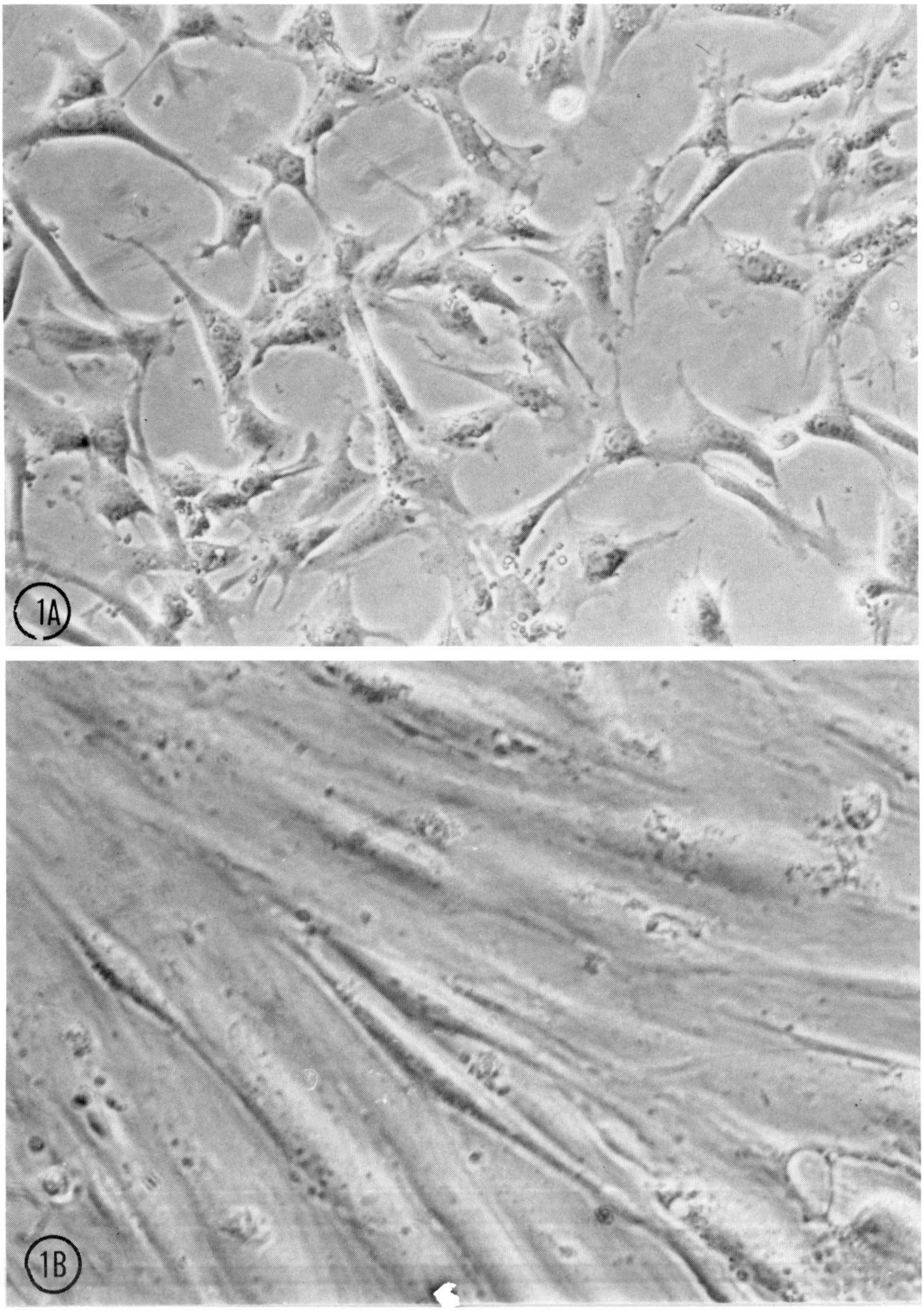

FIGURE 1. Phase-contrast micrograph comparing the morphological characteristics of bovine aortic SMC in the proliferating stage of growth (A) and the confluent growth stage (B). The latter displays the hill-and-valley, multilayered growth characteristic of SMC cultures.

Table 1
THE INCORPORATION OF [^{3}H] THYMIDINE, [^{3}H] PROLINE, AND [^{3}H] MANNOSE INTO COMPONENTS SYNTHESIZED BY PROLIFERATING AND CONFLUENT SMC IN CULTURE

	A Thymidine (dpm × 10/ mg protein)	B Proline (dpm × 10/ mg protein)	C Mannose (dpm × 10/ mg protein)
Proliferating SMC			
Medium		5.0	1.89
Cell layer	1.6	0.75	5.4
Insoluble matrix		0.58	2.08
Confluent SMC			
Medium		4.5	1.4
Cell layer	0.4	0.68	2.2
Insoluble matrix		0.41	0.86

Note: Mean values based on the level of incorporation in three identical flasks.

Table 2
SIZE DIFFERENCES[a] IN THE OLIGOMANNOSYL CORES DERIVED FROM ENDOGLYCOSIDASE TREATMENT OF CLYCOPEPTIDES FROM THE EXTRACELLULAR MATRIX OF SMC IN DIFFERENT PHASES OF GROWTH

	Endo H-sensitive (carbohydrate size)			Endo D-sensitive (carbohydrate size)		
Glycopeptide preparation	8—10	4—6	>10[b]	8—10	4—6	>10
Proliferating cells	24%	10%	66%	5%	30%	65%
Confluent cells	11%	19%	70%	15%	10%	75%

[a] Based on chromatographic analyses on Bio Gel P 4.[37]
[b] The oligosaccharides resistant to the endo cleavage were grouped in the fraction designated 10 carbohydrates.

higher amounts of large-core, endo H-sensitive oligos, and confluent cells have higher amounts of large-core, endo D-sensitive oligos.

The glycoproteins of the matrix from both phases of cell growth were radiolabeled with Man and separated on the basis of molecular weight. The overall pattern of glycoproteins synthesized by both phases of growth was similar. Profiles representative of either culture type are presented in Figure 2. Matrix components separated by polyacrylamide gel electrophoresis are presented in panel A (Figure 2), and similar preparations analyzed on gel permeation columns are shown in panel B (Figure 2). Several glycoproteins were observed by both methods ranging in molecular weight from under 20 kdaltons (band 10, panel A: peak at fraction 42, panel B) to over 500 kdaltons (band 1, panel A: peak at fraction 15, panel B). Whereas qualitatively the pattern of glycoproteins were similar in both phases of growth, the individual components differed with respect to their endo sensitivity.[37] The interstitial procollagens (Figure 2, panel A,

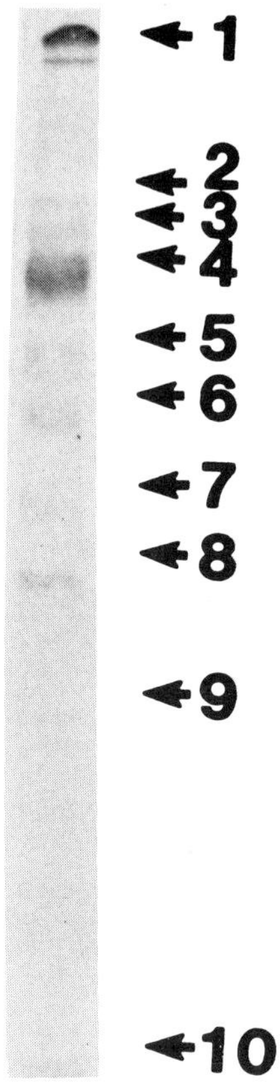

A

FIGURE 2. Slab gel electrophoresis of the Man-labeled glycoproteins associated with the insoluble matrix of SMC cultures. After deoxycholate treatment (0.5%) to remove soluble, cell-related components, the matrix was solubilized and analyzed on a 7.5% polyacrylamide slab gel. Glycoproteins were visualized by fluorography. At least ten bands of incorporation were evident and ranged in size from under 20 kdaltons (band 10) to over 500 kdaltons (band 1). The position of fibronectin (band 4) and the interstitial procollagens (band 5) were determined from the migration of standard glycoproteins under similar conditions. (B) Gel permeation chromatography of [^{3}H]-Man-labeled (open circles) and [^{35}S]-Met-labeled (closed circles) glycoproteins associated with the extracellular matrix of SMC. Components were analyzed on a 1.5 × 150-cm-column in a buffer containing 5 *M* urea. Molecular weight estimates were determined on the basis of elution volumes of glycoproteins with known molecular weights.

band 5) and fibronectin (Figure 2, panel A, band 4) were the Man-labeled components in these preparations. The other macromolecules corresponded to a class of components termed "structural glycoproteins".[12] Whereas the existence of Man-containing glycoproteins in the SMC matrix has been questioned by some,[44] our data support the concept of a distinct structural glycoprotein class in this context. Moreover, the results suggest that the transition from rapid growth to confluency (i.e., slow growth) in SMC cultures generally does not involve major qualitative changes in the extracellular glycoprotein. However, in accordance with previous studies employing other cell types,[33,34] our data support the contention that differences in the synthesis and/or processing of the oligos attached to components in the pericellular matrix are correlated with the state of growth of the SMC.[37]

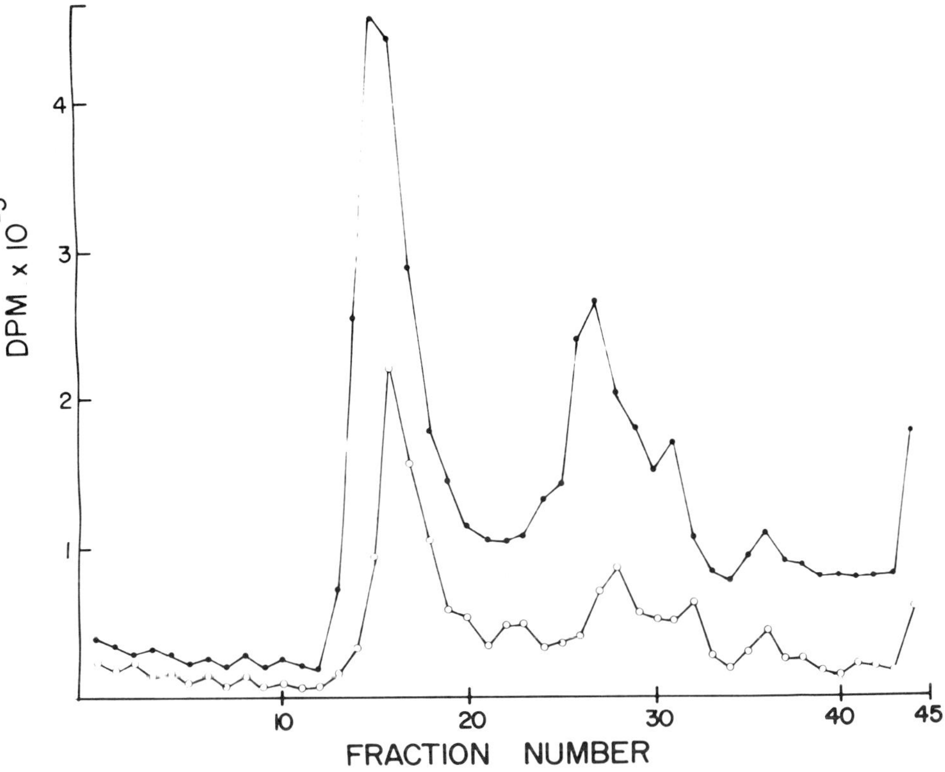

FIGURE 2B.

C. Intermediate and Interstitial Collagen Synthesis

Collagens associated with detergent extracts of the cellular layer or from the medium were routinely purified on DEAE cellulose to obtain interstitial and type V procollagen fractions using routine procedures.[23,45,46]

The simple oligos of many glycoproteins such as fibronectin[49] are converted to the complex type by the elimination of glucose and mannose residues. Subsequently, elongation takes place in the Golgi apparatus (with the addition of *N*-acetylglucosamine, galactose, sialic acid, and fucose) prior to the transportation to the cell membrane.[38,43] However, the data presented in Table 3 suggest that the oligos associated with the interstitial collagens (types I and III) are endo H-sensitive (simple or high mannose type) also after secretion from the cell, in analogy to observations made with other cell types.[20,47] In contrast to the interstitial procollagens the oligos attached to the type V procollagens were both endo H- and endo D-sensitive, suggesting that some processing had occurred already. Gel filtration (Figure 3) confirmed that endo D was primarily releasing oligos from these preparations, and all the products eluted as two major peaks in the region as expected for these carbohydrates.

Type V collagen may be deposited in the matrix as a procollagen without cleavage of the terminal extensions in a manner analogous to type IV.[19,50] Therefore, the oligos may play a role in determining the fate of the noncollagenous propeptides. The presence of endo D-sensitive oligos may prevent cleavage of the type V propeptides, while the simple oligos on the interstitial propeptides may "signal" their removal by the specific peptidases. Alternatively, the differences between the oligo content of the propeptides of the interstitial and type V collagens may be due to the presence of the unique P peptide that has been observed to be attached to type V procollagen during

Table 3
ENDOGLYCOSIDASE TREATMENT OF PROCOLLAGENS SECRETED BY SMC IN CULTURE

	Percent label released by Endo H		Percent label released by Endo D	
	Proliferating	Confluent	Proliferating	Confluent
Pro type V	48.9	40.2	25.6	31.2
Pro type I and III	59.0	45.6	4.6	5.2

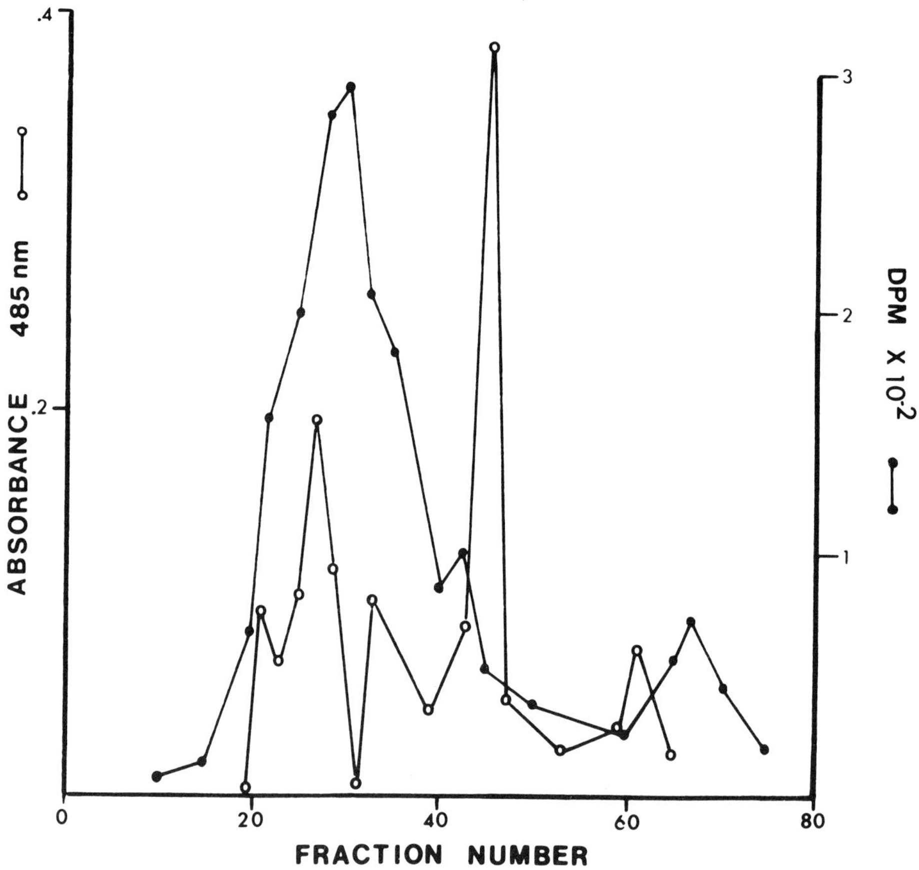

FIGURE 3. Gel filtration of endo D-cleaved oligos from type V procollagen. Solid circles represent label released, while the open circles depict the elution profile of pronase-derived carbohydrates from standard aortic type IV collagen. The peak at fraction 27 represents the oligo-sized components.

secretion.[50] Since differences were also observed in the levels of endo H- and endo D-sensitive oligos attached to type V between proliferating and confluent cultures (see Table 3), these components may be involved in determining the fate of this collagen during changes in the growth and interactions of cells. This collagen has been shown to play a role in cell proliferation and movement, and it is possible that its oligo composition is variable, pending the status of the cells.[51]

III. INTERACTION BETWEEN TYPE V COLLAGEN AND THE SMC SURFACE

A. Surface Glycoconjugates

At the ultrastructural level SMC in culture may be shown to be associated with a fairly well-defined BM (Figure 4). It was shown that the BM consists of a relatively lucent inner zone adjacent to the cell membrane, a poorly defined midzonal fibrillar network, and an outermost dense or compact zone. Fine discrete fibrils often are present within and account for the filamentous nature of the BM when it is cut tangentially.[52] Immunohistochemical observations suggest that these fibrils may be primarily type V collagen, which forms an extensive pericellular coat over the outer surface of the BM adjacent to the interstitial matrix.[53-55] It was also shown by immunohistochemical methods that this collagen was distributed on the surface of the SMC in such a way as to form a three-dimensional lattice anchoring the cell to the interstitial collagen matrix. Therefore, type V collagen seems to serve as a binding or connecting collagen linking the interstitial collagen fibers to the structures of the BM.[22,25]

Type V collagen covers the cell surface penetrating the BM at sites where it attaches to the plasma membrane at specific glycoconjugate receptors.[56] This form of direct-membrane interaction appears to be a unique property of the type V collagen since other classes of collagen interact with the cell surface by means of glycoprotein intermediaries, i.e., fibronectin for the interstitial and laminin for the BM collagens.[16,25] In addition to these two well characterized glycoprotein classes, several other polypeptides are thought to be involved in cell-collagen interactions, but the exact nature of these substances is not known.[57-61] The direct binding of type V collagen to the plasma membrane suggests that the biological function of this collagen differs, at least in some aspects, from that of the other collagens.

Deducing from the above, it was reasonable to assume that several independent systems were involved in SMC interaction with any one collagen. Therefore, several components at the surface of SMC involved in attachment to type V collagen have been investigated in our laboratories.[62] Our studies suggest that at least two surface glycoproteins on the SMC are either involved in binding type V directly or in a "cooperative" fashion in conjunction with a glycolipid. These studies were carried out according to procedures published by other workers.[56] Table 4 shows that the primary type V collagen-attachment system at the surface of SMC is probably a glycolipid since it was neuraminidase-sensitive and occurred even when protein synthesis was inhibited. However, in our system slightly elevated concentrations of cycloheximide (20 $\mu g/m\ell$ compared to 10 $\mu g/m\ell$)[56] caused a significant decrease in the final numbers of cells attached to type V collagen (approximately 25%). Moreover, the treatment with endo H resulted in a 20% decrease in binding, whereas that with endo D had no effect.

The nature of this cycloheximide-dependent system was determined by electrophoretic analyses of preparations of crude plasma membranes from radiolabeled cultures [^{35}S] methionine of [^{3}H] Man). Characteristic densitometric profiles are presented in Figure 5. Glycoprotein synthesis was compared in cells which were labeled in the process of attaching to a type V substrate and those grown on plastic (Figure 5, panels A and D, respectively). In general, the patterns of the major methionine (panels A and D) and Man-labeled (panels B and E) glycoproteins from the two culture types were comparable both in terms of numbers and positions (compare peaks 1 to 10 in panels A and D and peaks 1′ to 12′ in panels B and E). The components within the Man-labeled preparations that selectively bound to affinity columns composed of type V collagen attached to Sepharose were analyzed under similar conditions (Figure 5, panels C and E). Four peaks corresponding to glycoproteins were isolated from both culture types and corresponded to molecular weights of 50, 80, 180, and 200 kdaltons.

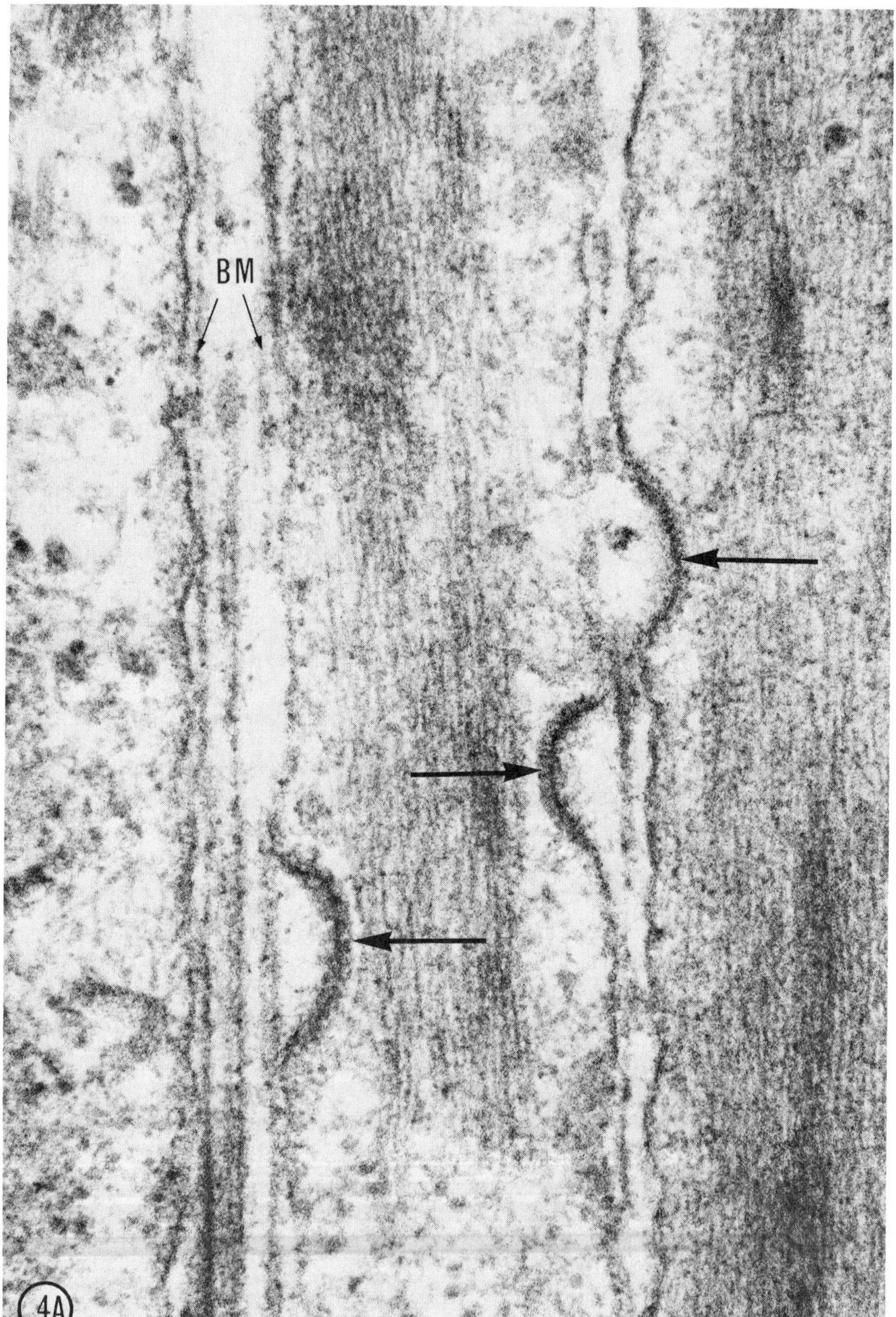

FIGURE 4A. Electron micrograph of a section through the "hill" of an early confluent (stationary) phase of bovine SMC in culture (ascorbate untreated). Complete or incomplete basement membranes (BM, small arrows) are adjacent to each of the three cells. Prominent Porter pits (large arrows) "open" in two of three instances, opposite a segment of a basement membrane. (Magnification × 100,000.)

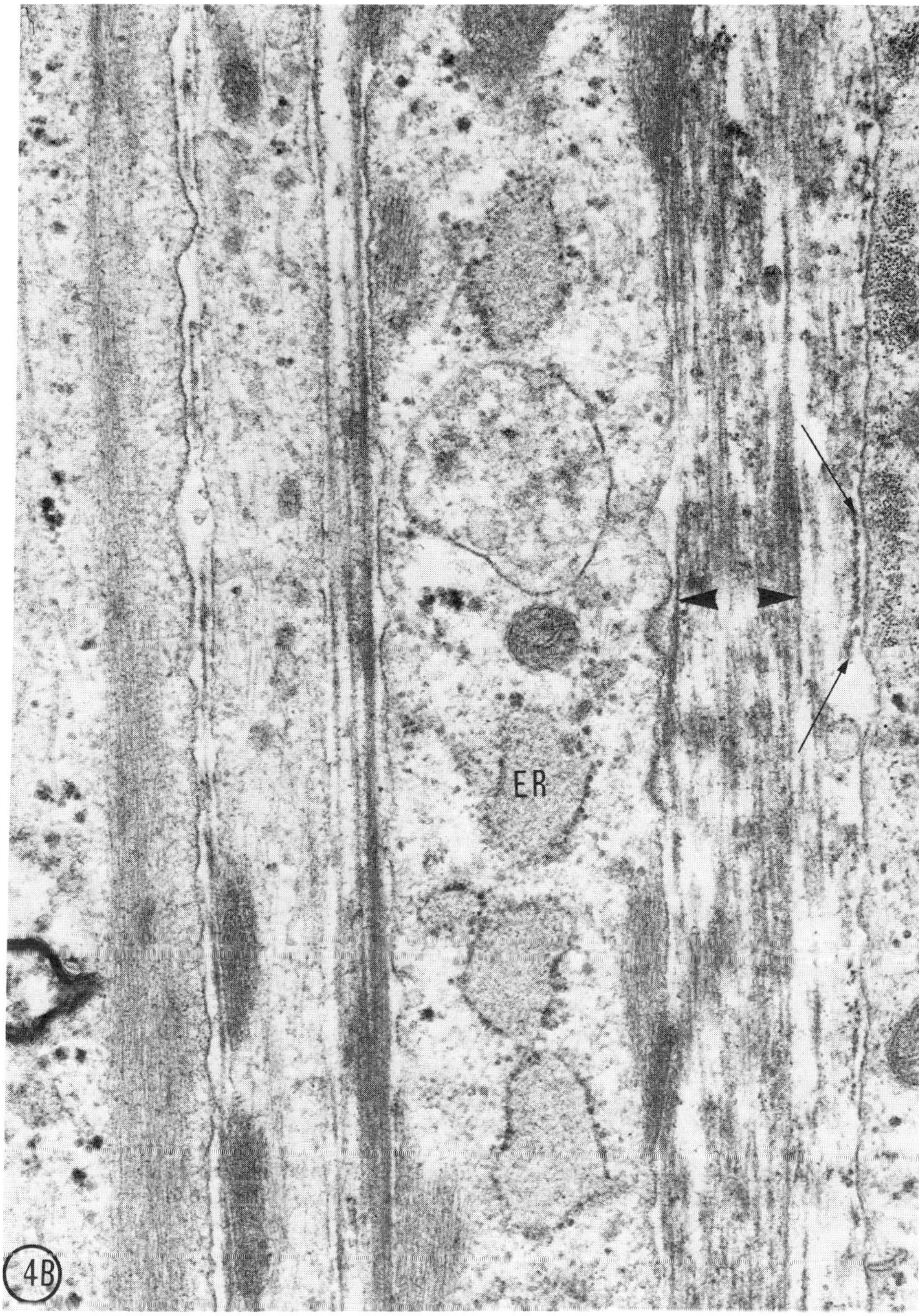

FIGURE 4B. Tissues as in Figure 4A but at a later stage in culture. The parallel-arranged SMC are in various stages of development, some containing longitudinal tracts (extreme left) or peripheral densities (=aggregates) (second cell from left) of myofilaments, whereas others are largely secretory with dilated profiles of rough surface endoplasmic reticulum (ER). Note that the intercellular space contains now more abundant basement membrane-like substance than that depicted in Figure 4A, but it is uncertain whether the ''fibrils'' in the space indicated (between two arrowheads) represent extracellular fibrils or remnants of (?tangential section through) a cell. Note that some segments of basement membranes appear ''beaded'' as if reflecting a cross-section through its constituent fibrils (between two arrows). (Magnification × 38,000.)

Table 4
THE EFFECT OF VARIOUS PRETREATMENTS ON BOVINE AORTIC SMC BINDING TO TYPE V COLLAGEN SUBSTRATES

Pretreatment[a]	Percentage[b] cells attached
None	76 ± 3.0
PBS	72 ± 4.0
Endo H	53 ± 2.6
Neuraminidase	31 ± 5.1
Endo H + neuraminidase	12 ± 3.8
Endo D	68 ± 4.8
Endo D + neuraminidase	33 ± 3.5
Cycloheximide[c] (20 μg/mℓ)	51 ± 4.4

[a] All enzymatic treatments were carried out immediately before attachment assay, as outlined.[62].

[b] The means ± SEM were computed from four independent determinations.

[c] Cycloheximide was present after trypsin/EDTA treatment and during attachment testing. This value is significantly smaller than that for nontreated cells with a *p*-value of 0.001 as determined by student's *t*-test.

The level of label associated with the 50- and 80-kdalton glycoproteins isolated from attaching cultures was consistently greater than that observed in cells grown independent of a type V collagen substrate. These type V collagen-binding components were also analyzed by gel permeation high-performance liquid chromatography (HPLC) (Figure 6). The 50- and 80-kdalton peaks both contained methionine and mannose label, had an affinity for concanavalin A, and were endo H-sensitive.

One of the type V collagen-binding glycoprotein fractions was similar in size (Figure 6, fraction 12, 200 kdaltons) to fibronectin, reacted to antibody against this protein, and had an ability to bind to several different collagens. Fibronectin has been shown to have this affinity for collagens, including the type V.[63] It could be expected to be present in the extracts reported herein primarily because it serves as the element of attachment for the SMC interaction with the interstitial collagens.[56] Some of the material in the 180 kdaltons (Figure 6, fraction 15) reacted to fibronectin antibody, but most of it was thought to be related to thrombospondin, which also has an affinity for type V collagen and is produced by SMC in culture.[64]

The 50-kdalton component had several features in common with another group of plasma membrane-associated glycoproteins which have been implicated in endodermal cell binding to collagen.[61] This class of molecules contained a 47-kdalton glycoprotein which not only shared a similarity to the 50-kdalton species because of its size, but also contained endo H-sensitive carbohydrates and selectively attached to a native undenatured type IV collagen substrate.[61] This 47-kdalton component was termed "colligin,"[61] and our studies suggest that molecules of this size may be common mediators of cell-collagen interactions. Support for this contention is provided by the findings that platelets attach themselves to type I collagen by means of a 65-kdalton species.[60] Moreover, a 47-kdalton, gelatin-binding component has been isolated from serum,[65] and a polypeptide of this size has been observed attached to fibronectin in hamster

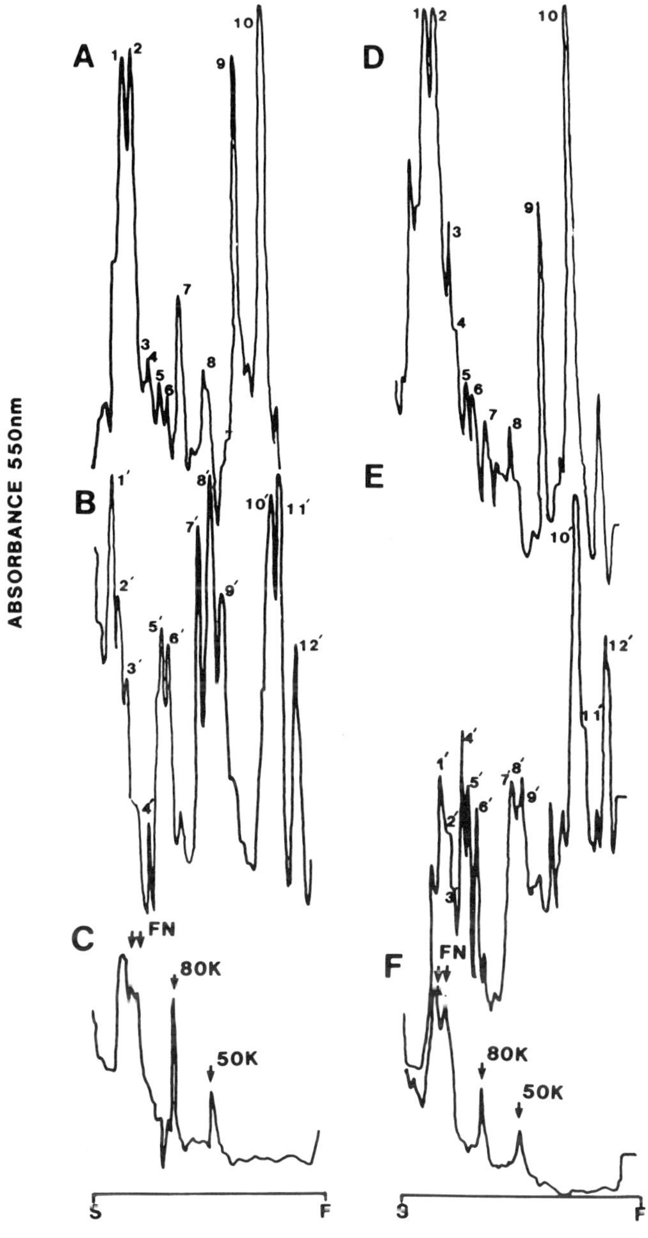

FIGURE 5. Slab gel electrophoresis of putative plasma membrane proteins from SMC labeled with [^{35}S] Met or [^{3}H] Man. After electrophoresis, the proteins were visualized by fluorography and subsequently scanned using an EC densitometer at 550 nm. Representative profiles obtained from cells attached to type V collagen are presented in panels A (Met-labeled, peaks 1 to 10) and B (Man-labeled, peaks 1′ to 12′). For comparison, similar preparations from SMC that had grown independent of type V substrates are presented in panels D (Met-labeled, peaks 1 to 10) and E (Man-labeled, peaks 1′ to 12′). Panels C and F represent the electrophoretic profiles of components contained in the preparations shown in panels A and D which were selectively retained by type V-Sepharose. The positions of fibronectin in panels C and F (arrows, FN), 80- and 50-kdalton glycoproteins, are indicated. The start (S) and dye front (F) of the gels are also indicated. Reproduced with permission of the National Research Council of Canada (re: Ref. 62).

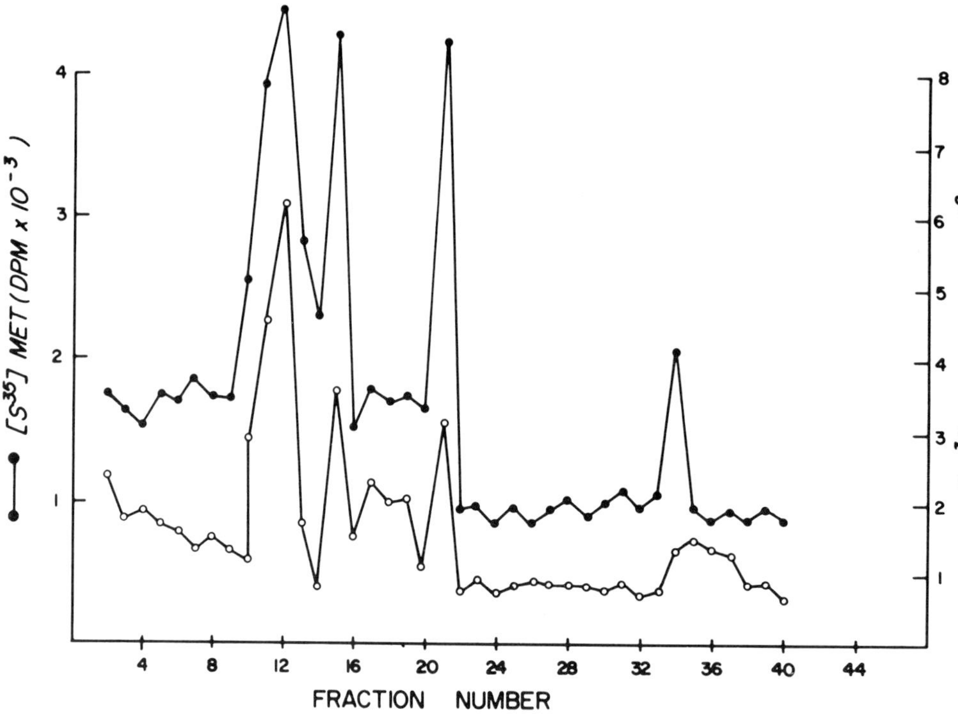

FIGURE 6. HPLC-elution profile of SMC-membrane glycoproteins selectively retained by affinity chromatography on type V collagen Sepharose shows that Met-labeled (closed circles) and Man-labeled (open circles) substances are both present in a given fraction. Molecular weight estimates based on the elution volumes of standard proteins indicated that the peaks at fractions 12, 15, and 21 correspond to molecular weights of 200, 180, and 50 kdaltons, respectively. The material eluting between fractions 16 and 20 had a mean molecular weight of 80 kdaltons (fraction 18). Reproduced with permission of the National Research Council of Canada (re: Ref. 62).

BHK cells.[66] Components similar in size to the 80-kdalton species have also been implicated in the binding of other cell types to collagen.[58,59]

Type V collagen consists of a helix composed of two dissimilar chains: (1) two $\alpha1$(V) chains and (2) one $\alpha2$(V) chain.[23] Previously, the type V$\alpha2$ chain was shown to contain the attachment region for SMC.[56] In our studies, cyanogen bromide peptides from radiolabeled type V chains were tested for their ability to bind to and promote SMC adhesion (Figure 7). They showed that the binding region is probably within the cyanogen bromide fragment 6 of this chain.[67,68] This fragment has a nonspecific amino acid composition which is a feature of these cell-attaching fragments such as the αCB 7 peptide of the $\alpha1$(I) chain.[69] Direct comparison between these fragments is difficult because the latter is almost three times the size of type V, $\alpha2$, CB 6.

B. Function Significance of Type V Collagen — A Role in Force Transference?

As outlined above, type V collagen has a unique distribution in the extracellular matrix,[22,53] especially at the surface of SMC and interacts with the cell with two distinct attachment systems.[62] These data suggest that its function in the matrix may be more than strictly of a structural nature.[70] A "collagenous net" has been suggested to play a role in the force transference of SMC,[71] and its existence was confirmed by ultrastructural studies.[72,73] According to this hypothesis, an SMC, after stimulation, actively increases in diameter, placing tension on the surrounding interstitial collagen fibers (in the same manner that the Chinese finger trap works). The stressed connective tissue

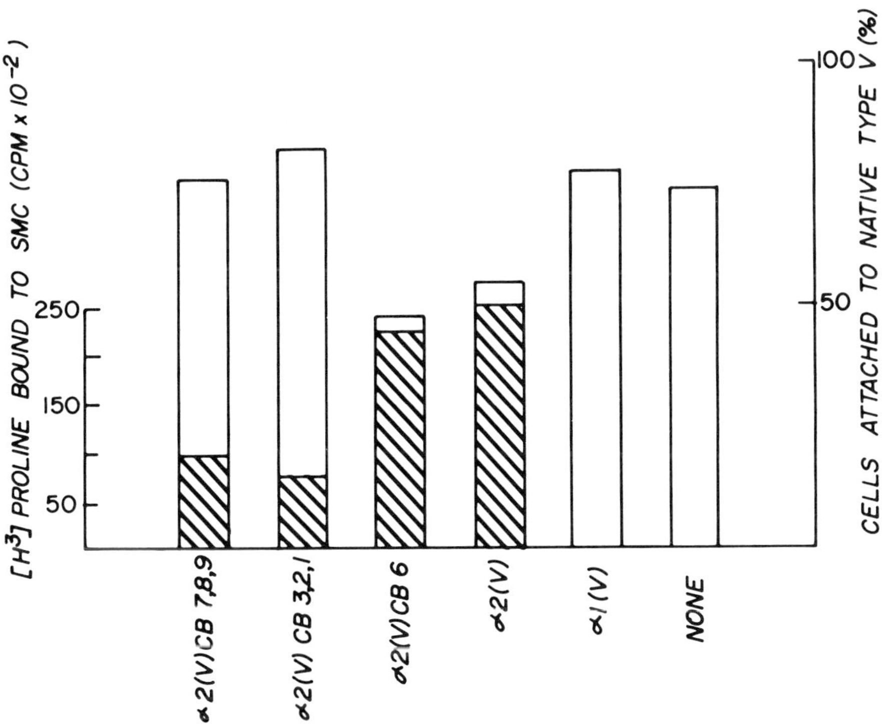

FIGURE 7. Histogram showing the percentage cells attached to type V collagen in the presence of collagen α chains and peptides from the same collagen (open bars). When these test components were radiolabeled, the amount of radioactivity associated with the cells was determined, as indicated by the cross-hatched bars. (From Leushner, J. R. A. and Haust, M. D., *Can. J. Biochem. Cell. Biol.*, in press, 1985. Reproduced with permission of the National Research Council of Canada.)

transmits force to the load. It is unlikely that the interstitial collagen fibers of types I and III[74,75] play an active role in this system. These latter fibers possess the characteristic properties generally ascribed to collagen; namely, high tensile strength, lack of extensibility, and structural stability. Therefore, the interstitial collagens probably play a relatively passive structural role in the process. On the other hand, data suggest that type V collagen is a major structural component of this "collagenous net". Since the arrangement of type V collagen may be envisaged as a latticework of pliable filaments, the original designation of this collagen as an exoskeleton[52] does not seem appropriate. This term may be better applied to the encompassing fibers of the interstitial collagens.

IV. TYPE VI COLLAGEN? A COMMON DENOMINATOR FOR COLLAGEN AND ELASTIC TISSUE

It was demonstrated that focal, tight, and relatively inflexible binding sites exist between SMC and the elastic laminae.[72,73] These appear to assure overall coordination of configurational changes between cell groups and elastic systems and to prevent slippage of cell layers. The material involved at these binding sites is probably a BM collagen. This was deduced from ultrastructural studies,[76,77] which have shown a spatial relation between the developing connective tissues and the BM of the SMC. In the original study,[76] elastic units were often observed to be attached to the BM of the SMC, and this appeared to be similar to that of the BM. Type III collagen has been shown to

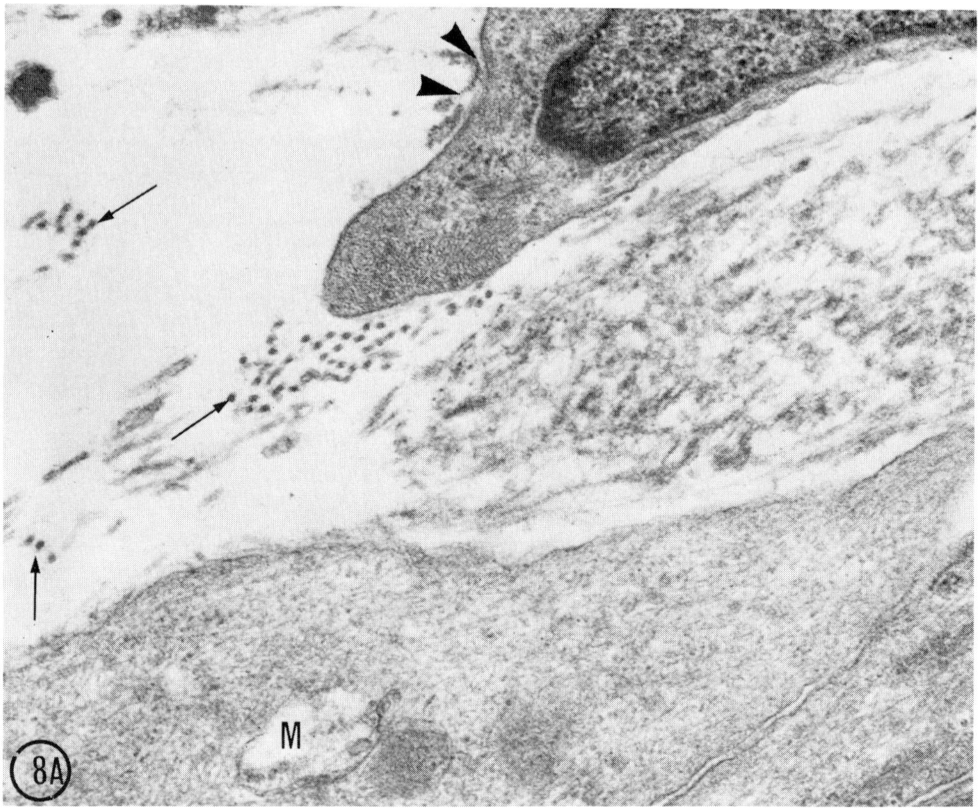

A

FIGURE 8(A). Electron micrograph of proliferative phase of SMC in culture with added ascorbate. Collagen fibrils (arrows) intermingled with fine and short fibrils are present in the intercellular space. Note the "beaded" short segment (between arrowheads) of the basement membrane of uppermost SMC. (M) poorly preserved mitochondrion. (Magnification × 30,000.) (B) Tissues as in Figure 8A. The collagen "bundle", indicated by an arrow, developed in close vicinity to an invagination of plasma membrane (of the SMC). Note that plasma membrane and the immediately adjacent cytoplasm opposite the developing collagen "bundle" is electron-opaque. The collagen fibrils are of varying and small diameter. The basement membrane "trailing" the cell (upper right corner, between the two arrowheads) has a "beaded" appearance. Much of this cell is otherwise not surrounded by a basement membrane. (Magnification × 90,000.)

be condensed at these attachment sites.[78] It is conceivable that type V filaments also terminate at these locations. The exact nature of the interaction of the cell with the elastin complexes remains unknown, but recent evidence has implicated a member of the intermediate class of collagens, particularly the type VI collagen, in this interaction.[79,80] Our observations suggest that in culture, type VI collagen is secreted in a form that shares characteristics with both type VI collagen and those of the aortic microfibrils.[24a,81]

A. Ultrastructural Analysis

In the absence of ascorbic acid, SMC secrete decreased amounts of the interstitial procollagens, while the production of glycoproteins (such as the microfibrillar component) increases.[26,81] At the ultrastructural level, ascorbate-treated (Figure 8) and untreated (Figure 9) cultures displayed marked differences. In the presence of ascorbate,

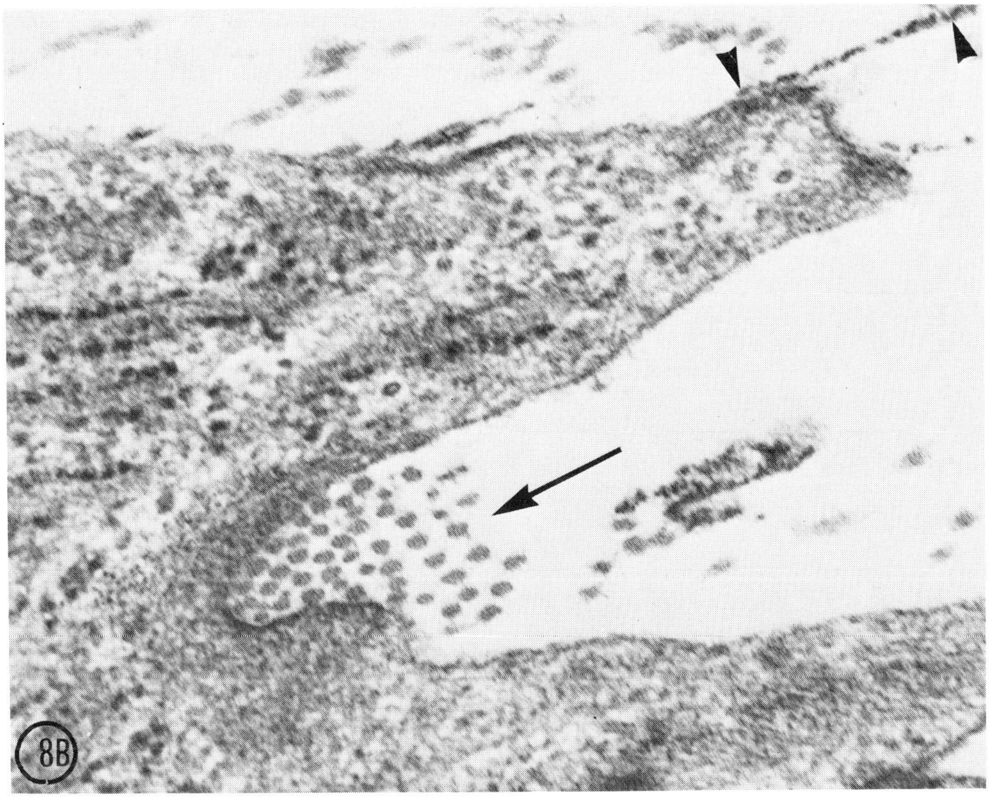

FIGURE 8B.

the extracellular spaces contained a considerable number of collagen fibrils (Figure 8), whereas, in its absence, microfibrillar (Figure 9) components dominated the field, and only a few collagen fibrils were observed on occasion.

B. Biochemical Analysis

Biochemical studies revealed that the secretion of the interstitial collagens decreased from 30 to 11% when ascorbate was omitted from the labeling media.[81] On the other hand, there was a concomitant increase in the more soluble collagenase-sensitive components in the 50% ammonium sulfate fraction in the absence of ascorbate. Several high-molecular-weight components could be demonstrated in this 50% fraction. Whereas the biochemical characteristics of type VI collagen have not been determined with certainty,[83,84] this collagen has several unique distinguishing features that are shared by the substance precipitating in the 50% ammonium sulfate fractions. These include:

1. It has a higher solubility than that of the other collagen types.[82,83]
2. In nonreduced form, it consists of high-molecular-weight aggregates, which, upon reduction, give rise to several low-molecular-weight polypeptides (from 30 to 70 kdaltons).[83]
3. Recent studies on a soluble form of this collagen suggest that these commonly found smaller chains may be derived from a 140-kdalton precursor.[85]

Our studies confirm that a 140-kdalton chain is associated with type VI collagen. They also indicate that it is very susceptible to proteolytic cleavage, being rapidly con-

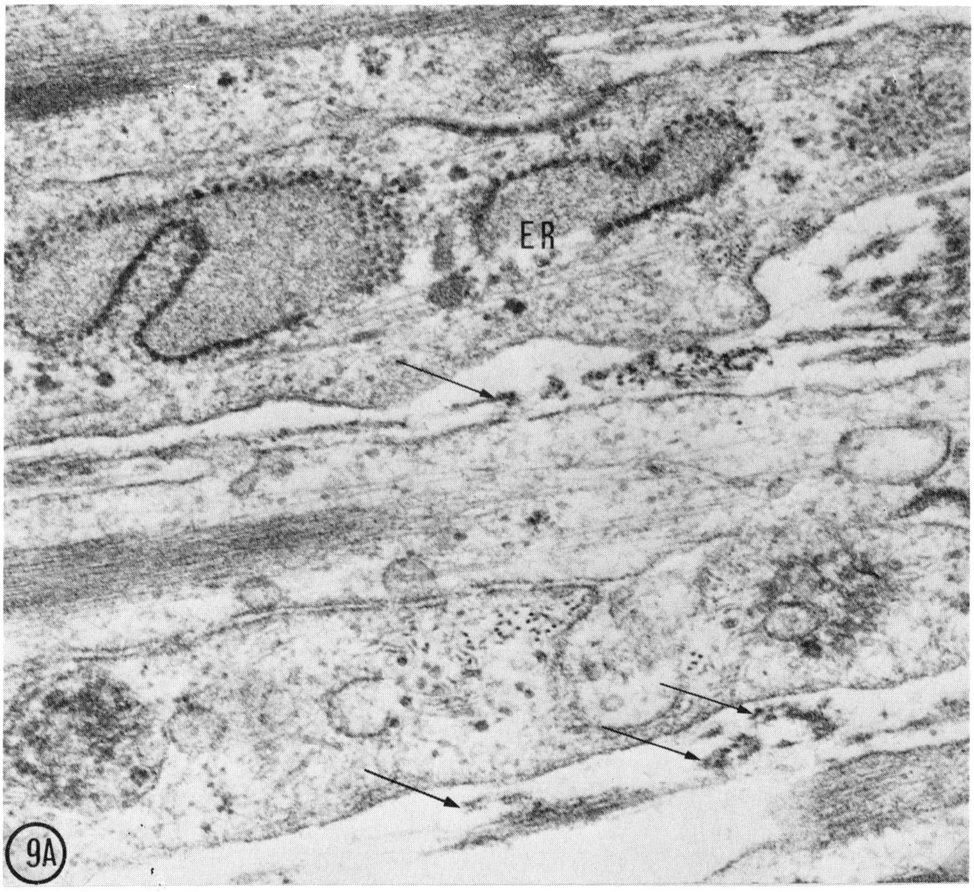

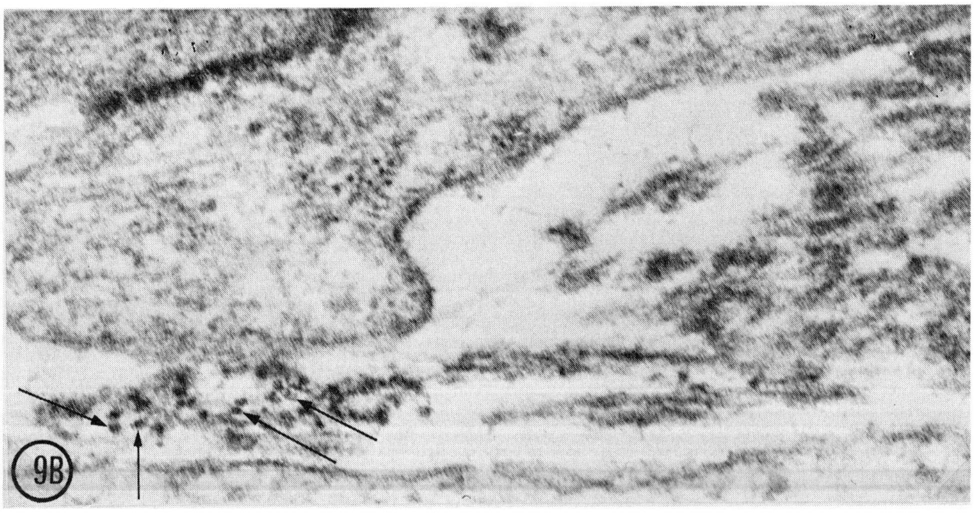

FIGURE 9(A). Electron micrograph of a ''hill'' of the confluent stationary phase of SMC in culture ''untreated'' with ascorbate. The cells are in various stages of differentiation and maturation, and some are in a distinctly secretory phase, with dilated profiles of endoplasmic reticulum (ER). No collagen fibrils are present. At this early stage, the intercellular space contains a modest number of microfibrils that appear to have a spatial relation to the cellular basement membrane (arrows). (Magnification × 40,000.) (B) Higher magnification of the area of and to the right indicated by the uppermost arrow in Figure 9B shows the microfibrils in cross- (arrows) and oblique (right half of photograph) cut. (Magnification × 80,000.)

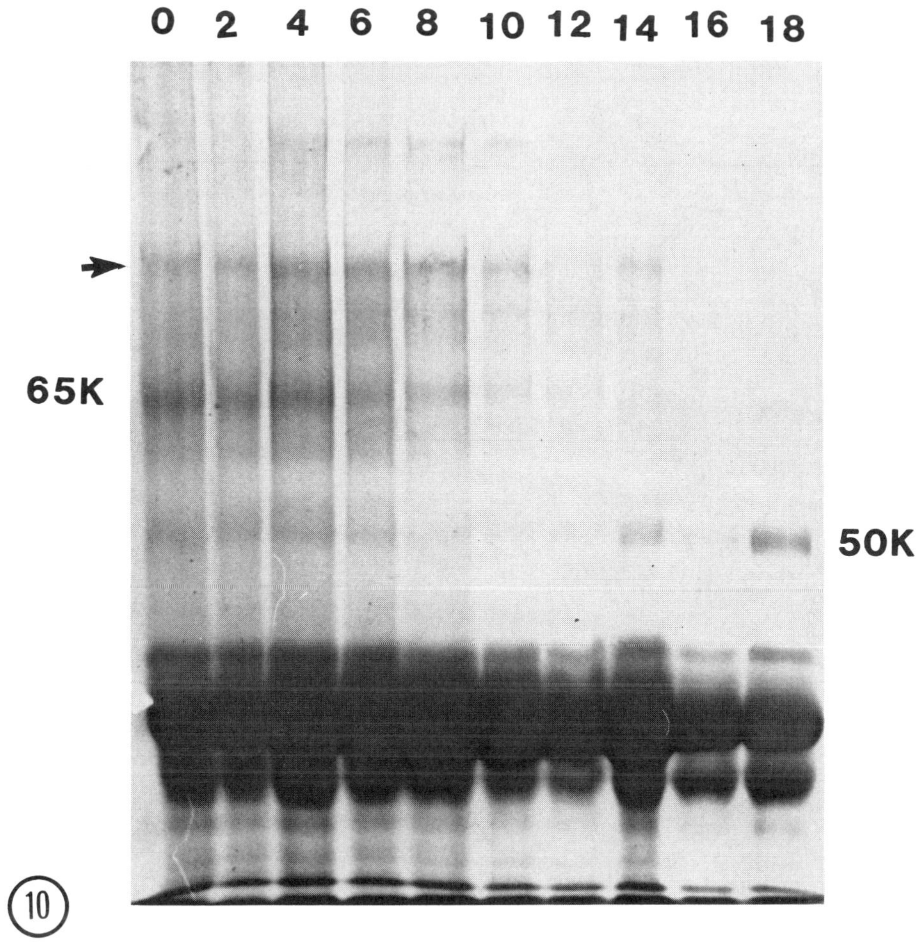

FIGURE 10. Slab gel electrophoresis of the 50% ammonium sulfate precipitate which had been incubated over an 18-hr interval in the absence of protease inhibitors. Aliquots of the sample were analyzed every 2 hr. The gradual disappearance of the 65- and 140-kdalton (arrow) chains with a concomitant increase in the 50-kdalton chain was apparent.

verted to smaller polypeptides in cultures. This is demonstrated by the experiment represented in Figure 10. The 50% ammonium sulfate precipitate (see previously) was incubated at room temperature in the absence of protease inhibitors for 16 hr, and aliquots were analyzed by electrophoresis at various times. A gradual decrease in the 140- (arrow) and 65-kdalton components and a concomitant increase in the smaller chains was evident with time. This protease sensitivity has been observed in previous studies on type VI collagen.[80]

It has been suggested once previously that low-molecular-weight collagens were produced by SMC in culture.[7] These polypeptides, designated CP 45, were assumed to be derived from BM and were present in equal proportions both in the medium and the cell layer. Whereas it is not possible to compare the components isolated in our own studies with the CP 45 because of limited information on the latter, it is tempting to postulate that they are the same or similar.

Other features of the collagen observed in our studies also indicate that it is type VI-like, e.g., the amino acid composition (see Table 5, column A). The high levels of hydroxyproline and hydroxylysine suggest that it is collagenous in nature, but features

Table 5
AMINO ACID COMPOSITION OF HIGH-MOLECULAR-WEIGHT COLLAGEN FROM SMC CULTURES[80,86]

Amino acid	A HMW-collagen	B Type VI	C Microfibril
4-HYP	37	39	26
ASP	92	102	102
THR	25	28	37
SER	38	47	45
GLU	102	93	117
PRO	93	76	64
GLY	227	243	179
ALA	80	51	69
CYS	30	23	57
VAL	36	37	18
MET	15	5	9
ILE	39	29	35
LEU	33	43	61
TYR	20	15	19
PHE	33	22	30
HYL	36	37	25
HIS	4	6	14
LYS	11	28	38
ARG	48	63	56

such as the low glycine levels and elevated amounts of half-cystine indicate noncollagenous sequences. These latter features are comparable to both: the published characteristics of type VI collagen (Table 5, column B) and the aortic microfibrillar component (Table 5, column C).

Recent electron microscopic examination of placental type VI collagen revealed its fibrillar nature,[79] which was similar to that of microfibrillar components of elastic tissue. These findings suggest that the type VI collagen and microfibrils of the extracellular space[24a] may be related. Moreover, the soluble form of microfibrils has been shown to be similar in size to the 140-kdalton type VI chain,[86,87] and our observations support this contention.

V. CLOSING COMMENTS

This report concerned recent work carried out in our laboratories on the isolation and characterization of glycoproteins and oligosaccharides associated with the extracellular matrix of bovine aortic SMC in culture. Two types of cell growth were distinguished during our studies: (1) a proliferative constantly growing phase and (2) a quiescent or confluent phase. Differences in composition and/or processing of oligosaccharides attached to the extracellular macromolecules were found depending upon the state of growth of the cell, including variation in size and sensitivity to endoglycosidases (endo). The endo H-sensitive oligomannosyl cores from the matrix of proliferating cells were larger than those associated with confluent cultures. On the other hand, the latter cells contained endo D-sensitive oligomannosyl residues which were larger than those from the matrix of proliferating cells. In the presence of ascorbic acid, procollagens and fibronectin were the major glycoproteins, while up to 11 other minor mannose-containing components were observed in the matrix secreted by both phases of the cell growth. Variation in the endo sensitivity was observed in several of the glycoproteins when the different cells were compared, especially with respect to the procollagens and fibronectin. Some of the variability in the procollagen fraction was ascribed to procollagen type I, which, unlike the interstitial types, contained endo D-sensitive components.

Data indicate that SMC interaction with type V collagen may involve both glycolipid and glycoprotein components. The latter are 50,000 and 80,000 daltons in size, contain simple oligosaccharides, and bind preferentially to the cyanogen bromide number 6 of the $\alpha 2(V)$ chain. The observations on type V collagen were interpreted in terms of a hypothetical "collagenous net" theory. Evidence suggested that type V collagen was the major structural component of this net, and therefore type V collagen may be involved in force transference in the SMC function.

In the absence of added ascorbic acid, the SMC cultures secreted reduced amounts of interstitial collagens but produced a collagenous high-molecular-weight polypeptide complex into the media. Reduction of disulfide cross-links within this complex gave rise to a series of low-molecular-weight subunits comparable in size and composition to type VI collagen and sharing some features with the aortic microfibrils. These data were confirmed and correlated with the ultrastructural features of the extracellular components of SMC cultures.

SMC in culture synthesize a heterogeneous group of carbohydrate-containing proteins. The exact role of these components remains uncertain, primarily because of their diversity. It is by complex associations, possibly involving their carbohydrate side chains, that these macromolecules form diverse but integrated matrices. These vary in composition from the plasma membrane to the cell surface and finally to the intercellular space. Compositional differences in these substances are believed to be associated with functional changes that occur in the cell during adhesion, proliferation, and migration. Therefore, changes in the amounts and types of both the polypeptide and carbohydrate components of these macromolecules may play an important role in the physiological and pathological changes that occur in aorta. This complex interplay is exemplified by our observations on the interaction of the cell with type V collagen, as well as the possible shared identity of type VI collagen and microfibrillar components associated with elastic elements. Moreover, the unique characteristics of these macromolecules have necessitated a complete reevaluation of our concepts that collagens serve strictly as structural elements of either an interstitial or BM distribution. On the contrary, as more data accumulate on such components, e.g., the intermediate collagens, their functional roles will prove as diverse as their biochemical characteristics.

ACKNOWLEDGMENTS

This work was supported by a grant-in-aid T.3-11 from the Ontario Heart and Stroke Foundation, Toronto, Canada. The authors wish to thank Mrs. M. John and Mr. R. Dewar for their skillful technical assistance and Mrs. B. Gardiner and Ms. J. Balint for their efficient typing of the manuscript.

REFERENCES

1. Haust, M. D. and More, R. H., Significance of the smooth muscle cell in atherogenesis, in *Evolution of the Atherosclerotic Plaque,* Jones, R. J., Ed., University of Chicago Press, Chicago, 1963.
2. Haust, M. D., Atherosclerosis and smooth muscle cells, in *Biochemistry of Smooth Muscles,* Vol. 2, Stephens, N. L., Ed., CRC Press, Boca Raton, Fla., 1983, 189.

3. Boissel, J. P., Bourdillon, M. C., Loire, R., and Grouzet, B., Histological arguments for collagen and elastin synthesis by primary cultures of rat aortic media cells, *Atherosclerosis,* 25, 107, 1976.
4. Daoud, A. S., Fritz, K. E., Singh, J., Augustyn, J. M., and Jarmolych, J., Arterial mesenchyme and arteriosclerosis. Production of monosaccharides, collagen and elastic tissue by aortic medial explants, *Adv. Exp. Med. Biol.,* 43, 281, 1974.
5. Jarmolych, J., Daoud, A. S., Landau, J., Fritz, K. E., and McElvene, E., Aortic media explants. Cell proliferation and production of mucopolysaccarides, collagen and elastic tissue, *Exp. Mol. Pathol.,* 9, 171, 1968.
6. Schwatrz, E., Adamany, A. M., and Blumenfeld, O. O., Extracellular proteins of the calf aortic media smooth muscle cells in culture, *Biochim. Biophys. Acta,* 624, 531, 1980.
7. Mayne, R., Vail, M. S., and Miller, E. J., Characterization of the collagen chains synthesized by cultured smooth muscle cells derived from rhesus monkey thoracic aorta, *Biochemistry,* 17, 446, 1978.
8. Rauterberg, J., Allam, S., Brehmer, U., Wirth, W., and Hauss, W. H., Characterization of the collagen synthesized by cultured human smooth muscle cells from fetal and adult aorta, *Hoppe Seyler's Z. Physiol. Chem.,* 358, 401, 1977.
9. Scott, D. M., Harwood, R., Grant, M. E., and Jackson, D. S., Characterization of the major collagen species present in porcine aortae and the synthesis of their precursors by smooth muscle cells in culture, *Connect. Tissue Res.,* 5, 7, 1977.
10. Layman, D. L. and Titus, J. L., Synthesis of type I collagen by human smooth muscle cells in vitro, *Lab. Invest.,* 33, 103, 1975.
11. De Clerck, Y. A. and Jones, P. A., The effect of ascorbic acid on the nature of production of collagen and elastin by rat smooth muscle cells, *Biochem. J.,* 186, 217, 1980.
12. Jones, P. A., Scott-Burden, T., and Gevers, W., Glycoprotein, elastin and collagen secretion by rat smooth muscle cells, *Proc. Natl. Acad. Sci. U.S.A.,* 76, 353, 1979.
13. Kefalides, N. A., Basement membranes: structural and biosynthetic considerations, *J. Invest. Dermatol.,* 65, 85, 1975.
14. Bornstein, P. and Sage, H., Structurally distinct collagen types, *Annu. Rev. Biochem.,* 49, 957, 1980.
15. Ruoslahti, E., Engvall, E., and Hayman, E. G., Fibronectin: current concepts of its structure and function, *Collagen Rel. Res.,* 1, 95, 1981.
16. Kleinman, H. K., Klebe, R. J., and Martin, G. R., Role of collagenous matrices in the adhesion and growth of cells, *J. Cell Biol.,* 88, 473, 1981.
17. Timpl, R., Rohde, H., Gehron-Robey, P., Rennard, S. I., Foidart, J-M., and Martin, G. R., Laminin — a glycoprotein from basement membranes, *J. Biol. Chem.,* 254, 9933, 1979.
18. Foidart, J-M., Bere, E. W., Yaar, M., Rennard, S. I., Gullino, M., Martin, G. R., and Katz, S. I., Distribution and immunoelectron microscopic localization of laminin, a noncollagenous basement membrane glycoprotein, *Lab. Invest.,* 42, 336, 1980.
19. Robert, A. M., Robert, B., and Robert, L., Chemical and physical properties of structural glycoproteins, in *The Chemistry and Molecular Biology of the Intercellular Matrix,* Balazs, I., Ed., Academic Press, New York, 1970.
20. Clark, C. C., The distribution and initial characterization of oligosaccharide units on the COOH-terminal propeptide extensions of the Pro- 1 and Pro- 2 chains of type I procollagen, *J. Biol. Chem.,* 254, 10798, 1979.
21. Levine, M. J. and Spiro, R. G., Isolation from glomerular basement membrane of a glycopeptide containing both asparagine-linked and hydroxylysine-linked carbohydrate units, *J. Biol. Chem.,* 254, 8121, 1979.
22. Konomi, H., Hayashi, T., Nakayasu, K., and Arima, M., Localization of type V collagen and type IV collagen in human cornea, lung, and skin, *Am. J. Pathol.,* 116, 417, 1984.
23. Leushner, J. R. A. and Haust, M. D., Characterization of basement membrane collagens of bovine aorta, *Atherosclerosis,* 50, 11, 1984.
24. Poste, G. and Nicolson, G. L., Eds., *Cytoskeletal Elements and Plasma Membrane Organization,* North-Holland, Amsterdam, 1981.
24a. Haust, M. D., Fine fibrils of extracellular space (microfibrils). Their structure and role in connective tissue organization, *Am. J. Pathol.,* 47, 1113, 1965.
25. Hedman, K., Kurkinen, M., Alitalo, K., Vaheri, A., Johansson, S., and Hook, M., Isolation of the pericellular matrix of human fibroblast cultures, *J. Cell Biol.,* 81, 83, 1979.
26. Schwartz, E., Bienkowski, R. S., Coltoff-Schiller, B., Goldfischer, S., and Blumenfeld, O. O., Changes in the components of extracellular matrix and in growth properties of cultured aortic smooth muscle cells upon ascorbate feeding, *J. Cell Biol.,* 92, 462, 1982.
27. Gospodarowicz, D., Hirabayashi, K., Giguere, L., and Tauber, J. P., Factors controlling the proliferative rate, final cell density, and final life span of bovine vascular smooth muscle cells in culture, *J. Cell Biol.,* 89, 568, 1981.

28. Poste, G. and Nicolson, G. L., *Cell Surface Reviews*, Vol. 5, Elsevier, Amsterdam, 1978.
29. Alpin, J. D. and Hughes, R. C., Complex carbohydrates of the extracellular matrix structures, interactions and biological roles, *Biochim. Biophys. Acta*, 694, 375, 1982.
30. Subtelny, S. and Wessels, N. K., *The Cell Surface: Mediator of Developmental Processes*, Academic Press, New York, 1980.
31. Buck, C. A., Glick, M. C., and Warren, L., Effect of growth on the glycoproteins from the surface of control and Rous sarcoma-virus transformed hamster cells, *Biochemistry*, 10, 2176, 1971.
32. Ceccarini, C., Muramatsu, T., Tsang, J., and Atkinson, P. H., Growth-dependent alterations in oligomannosyl cores of glycopeptides, *Proc. Natl. Acad. Sci. U.S.A.*, 72, 3139, 1975.
33. Muramatsu, T., Koide, N., Ceccarini, C., and Atkinson, P. H., Characterization of mannose-labelled glycopeptides from human diploid cells and their growth-dependent alterations, *J. Biol. Chem.*, 251, 4673, 1976.
34. Sasak, W., Herscovics, A., and Quaroni, A., Cell-density-dependent changes in cell-surface glycopeptides and in adhesion of cultured intestinal epithelial cells, *Biochem. J.*, 201, 359, 1982.
35. Dennis, J., Waller, C., Timpl, R., and Schirrmacher, V., Surface sialic acid reduces attachment of metastatic tumour cells to collagen type IV and fibronectin, *Nature (London)*, 300, 274, 1982.
36. Görög, P. and Born, G. V. R., Increased uptake of circulating low-density lipoproteins and fibrinogen in arterial walls after removal of sialic acids from their endothelial surface, *Br. J. Exp. Pathol.*, 63, 447, 1982.
37. Leushner, J. R. A. and Haust, M. D., Oligosaccharides of matrix glycoproteins synthesized by cultured bovine aortic smooth muscle cells during various phases of growth, submitted for publication, 1985.
38. Staneloni, R. J. and Leloir, L. F., The biosynthetic pathway of the asparagine-linked oligosaccharides of glycoproteins, *CRC Crit. Rev. Biochem.*, 12(4), 289, 1982.
39. Baenziger, J. U. and Fiete, D., Structural determinants of concanavalin A specificity for oligosaccharides, *J. Biol. Chem.*, 254, 2400, 1979.
40. Narasimhan, S., Wilson, J. R., Martin, E., and Schachter, H., A structural basis for four distinct elution profiles on Concanavalin-A-Sepharose affinity chromatography of glycopeptides, *Can. J. Biochem.*, 57, 83, 1979.
41. Arakawa, M. and Muramatsu, T., Endo-*N*-acetylglucosaminidases acting on the carbohydrate moieties of glycoproteins. The differential specificities of the enzymes from Streptomyces griseus and Diplococcus pneumoniae, *J. Biochem. (Tokyo)*, 76, 307, 1974.
42. Varki, A. and Kornfeld, S., The spectrum of anionic oligosaccharides release by endo-*N*-acetylglucosaminidase H from glycoproteins, *J. Biol. Chem.*, 258, 2808, 1983.
43. Kornfeld, S., Li, E., and Tabas, I., The synthesis of complex-type oligosaccharides, *J. Biol. Chem.*, 253, 7771, 1978.
44. Bach, P. R. and Bentley, J. P., Structural glycoprotein, fact or artefact, *Connect. Tissue Res.*, 7, 185, 1980.
45. Sodek, J. and Limeback, H. F., Comparison of the rates of synthesis, conversion, and maturation of type I and type III collagens in rat periodontal tissues, *J. Biol. Chem.*, 254, 10496, 1979.
46. Kumamoto, C. A. and Fessler, J. H., Biosynthesis of A,B procollagen, *Proc. Natl. Acad. Sci. U.S.A.*, 77, 6434, 1980.
47. Anttinen, H., Oikarinen, A., Ryhanen, L., and Kivirikko, K. I., Evidence for the transfer of mannose to the extension peptides of procollagen within the cisternae of the rough endoplasmic reticulum, *FEBS Lett.*, 87, 222, 1978.
48. Choi, M. G. and Hynes, R. O., Biosynthesis and processing of fibronectin in NIL.8 hamster cells, *J. Biol. Chem.*, 254, 12050, 1979.
49. Narayanan, A. S. and Page, R. C., Biosynthesis and regulation of type V collagen in diploid human fibroblasts, *J. Biol. Chem.*, 258, 11694, 1983.
50. Fessler, L. I., Kumamoto, C. A., Meis, M. E., and Fessler, J. H., Assembly and processing of procollagen V (AB) in chick blood vessels and other tissues, *J. Biol. Chem.*, 256, 9640, 1981.
51. Stenn, K. S., Madri, J. A., and Roll, F. J., Migrating epidermis produces AB_2 collagen and requires continual collagen synthesis for movement, *Nature (London)*, 277, 229, 1979.
52. Haust, M. D., Connective tissue in atherosclerosis, in *Atherosclerosis IV*, Schettler, G., Goto, Y., Hata, Y., and Kose, G., Eds., Springer-Verlag, New York, 1977.
53. Gay, S., Martinez-Hernandez, A., Rhodes, R. K., and Miller, E. J., The collagenous exocytoskeleton of smooth muscle cells, *Collagen Rel. Res.*, 1, 377, 1981.
54. Linsenmayer, T. F., Fitch, J. M., Schmid, T. M., Zak, N. B., Gibney, E., Sanderson, R. D., and Mayne, R., Monoclonal antibodies against chicken type V collagen: production, specificity, and use for immunocytochemical localization in embryonic cornea and other organs, *J. Cell Biol.*, 96, 124, 1983.
55. Martinez-Hernandez, A., Gay, S., and Miller, E. J., Ultrastructural localization of type V collagen in rat kidney, *J. Cell Biol.*, 92, 343, 1982.

56. Grotendorst, G. R., Seppa, H. E. J., Kleinman, H. K., and Martin, G. R., Attachment of smooth muscle cells to collagen and their migration toward platelet-derived growth factor, *Proc. Natl. Acad. Sci. U.S.A.*, 78, 3669, 1981.
57. Yamada, K. M., Cell surface interactions with extracellular materials, *Annu. Rev. Biochem.*, 52, 761, 1983.
58. Vartio, T. and Vaheri, A., A gelatin-binding 70,000-dalton glycoprotein synthesized distinctly from fibronectin by normal and malignant adherent cells, *J. Biol. Chem.*, 256, 13085, 1981.
59. Vartio, T., Hovi, T., and Vaheri, A., Human macrophages synthesize and secrete a major 95,000 dalton gelatin binding protein distinct from fibronectin, *J. Biol. Chem.*, 257, 8862, 1982.
60. Chiang, T. M. and Kang, A. H., Isolation and purification of collagen 1(I) receptor from human platelet membrane, *J. Biol. Chem.*, 257, 7581, 1982.
61. Kurkinen, M., Taylor, A., Garrels, J. I., and Hogan, B. L. M., Cell surface-associated proteins which bind native type IV collagen or gelatin, *J. Biol. Chem.*, 259, 5915, 1984.
62. Leushner, J. R. A. and Haust, M. D., Glycoproteins on the surface of smooth muscle cells involved with their interaction with type V collagen, *Can. J. Biochem. Cell Biol.*, in press, 1985.
63. Engvall, E., Ruoslahti, E., and Miller, E. J., Affinity of fibronectin to collagens of different genetic types and to fibrinogen, *J. Exp. Med.*, 147, 1584, 1978.
64. Mumby, S. M., Raugi, G. J., and Bornstein, P., Interactions of thrombospondin with extracellular matrix proteins: selective binding to type V collagen, *J. Cell Biol.*, 98, 646, 1984.
65. Isemura, M., Sato, N., and Yosizawa, Z., Isolation and characterization of glycine-rich gelatin binding protein; a plasma glycoprotein which interacts with gelatin, arginine, heparin and hyaluronic acid, *J. Biol. Chem.*, 257, 14854, 1982.
66. Hughes, R. C., Butters, T. D., and Aplin, J. D., Cell surface molecules involved in fibronectin-mediated adhesion. A study using specific antisera, *Eur. J. Cell Biol.*, 26, 198, 1981.
67. Rhodes, R. K., Gibson, K. D., and Miller, E. J., Isolation and characterization of the cyanogen bromide peptides derived from the human $\alpha 2$(V) collagen chain, *Biochemistry*, 20, 3117, 1981.
68. Miller, E. J., Rhodes, R. K., and Furuto, D. K., Identification of collagen chains as a function of cyanogen bromide peptide patterns using gel permeation high-performance liquid chromatography, *Collagen Rel. Res. Clin. Exp.*, 3, 79, 1983.
69. Kleinman, H. K., McGoodwin, E. B., Martin, G. R., Klebe, R. J., Fietzek, P. P., and Woolley, D. E., Localization of the binding site for cell attachment in the α(I) chain of collagen, *J. Biol. Chem.*, 253, 5642, 1978.
70. Leushner, J. R. A., Type V collagen and the transmission of forces in smooth muscle cell function: Dupuytren's disease as a model system, *J. Inferential Deductive Biol.*, in press, 1985.
71. Mullins, G. L. and Guntheroth, W. G., A collagen net hypothesis for force transference of smooth muscle, *Nature (London)*, 206, 592, 1965.
72. Wolinsky, H. and Glagov, S., Structural basis for the static mechanical properties of the aortic media, *Circ. Res.*, 14, 400, 1964.
73. Clark, J. M. and Glagov, S., Structural integration of the arterial wall. I. Relationships and attachments of the medial smooth muscle cells in normally distended and hyperdistended aortas, *Lab. Invest.*, 40, 587, 1979.
74. Trelstad, R. L., Human aorta collagens: evidence for three distinct species, *Biochem. Biophys. Res. Commun.*, 57, 717, 1974.
75. Chung, E. and Miller, E. J., Collagen polymorphism: characterization of molecules with the chain composition $[\alpha 1(III)]_3$ in human tissues, *Science*, 183, 1200, 1974.
76. Haust, M. D., More, R. H., Benscome, S. A., and Balis, J. U., Elastogenesis in human aorta: an electron microscopic study, *Exp. Mol. Pathol.*, 4, 508, 1965.
77. Leung, D. Y. M., Glagov, S., and Mathews, M. B., Elastin and collagen accumulation in rabbit ascending aorta and pulmonary trunk during postnatal growth. Correlation of cellular synthetic response with medial tension, *Circ. Res.*, 41, 316, 1977.
78. Gay, S., Balleisen, L., Remberger, K., Fietzek, P. P., Adelmann, B. C., and Kuhn, K., Immunohistochemical evidence for the presence of collagen type III in human arterial walls, arterial thrombi, and in leucocytes incubated with collagen in vitro, *Klin. Wochenschr.*, 53, 899, 1975.
79. Furthmayr, H., Widemann, H., Timpl, R., Odermatt, E., and Engel, J., Electron-microscopical approach to a structural model of intima collagen, *Biochem. J.*, 211, 303, 1983.
80. Odermatt, E., Risteli, J., VanDeldin, V., and Timpl, R., Structural diversity and domain composition of a unique collagenous fragment (intima collagen) obtained from human placenta, *Biochem. J.*, 211, 295, 1983.
81. Leushner, J. R. A. and Haust, M. D., The effect of ascorbate on the synthesis of the minor (non-interstitial) collagens by cultured bovine aortic smooth muscle cells, *Biochim. Biophys. Acta.*, in press, 1986.
82. Chung, E., Rhodes, R. K., and Miller, E. J., Isolation of three collagenous components of probable basement membrane origin from several tissues, *Biochem. Biophys. Res. Commun.*, 71, 1167, 1976.

83. Furuto, D. K. and Miller, E. J., Isolation of a unique collagenous fraction from limited pepsin digests of human placental tissue, *J. Biol. Chem.*, 255, 290, 1980.
84. Jander, R., Rauterberg, J., Voss, B., and vonBassewitz, D. B., A cysteine-rich collagenous protein from bovine placenta, *Eur. J. Biochem.*, 114, 17, 1981.
85. Jander, R., Troyer, D., and Rauterberg, J., A collagen-like glycoprotein of the extracellular matrix is the undegraded form of type VI collagen, *Biochemistry*, 23, 3675, 1984.
86. Gibson, M. A. and Cleary, E. G., A collagen-like glycoprotein from elastin-rich tissues, *Biochem. Biophys. Res. Commun.*, 105, 1288, 1982.
87. Sear, C. H. J., Grant, M. E., and Jackson, D. S., The nature of the microfibrillar glycoprotein of elastic fibers, *Biochem. J.*, 194, 587, 1981.

INDEX

D

E

F

G

H

N

O

P

R

S

T

U

V

W